Technology and (Dis)Empowerment

Technology and (Dis)Empowerment: A Call to Technologists

AADITESHWAR SETH

Indian Institute of Technology Delhi and Gram Vaani Community Media

emerald PUBLISHING

United Kingdom – North America – Japan – India – Malaysia – China

Emerald Publishing Limited
Howard House, Wagon Lane, Bingley BD16 1WA, UK

First edition 2022

British Library Cataloguing in Publication Data
A catalogue record for this book is available from the British Library

ISBN: 978-1-80382-394-2 (Print)
ISBN: 978-1-80382-393-5 (Online)
ISBN: 978-1-80382-395-9 (Epub)

Printed and bound by CPI Group (UK) Ltd, Croydon, CR0 4YY

ISOQAR certified Management System, awarded to Emerald for adherence to Environmental standard ISO 14001:2004.

ISOQAR REGISTERED

Certificate Number 1985
ISO 14001

INVESTOR IN PEOPLE

Endorsements

If you want to use information technology to make a positive difference in the world, then you need to read this book. Aadi Seth combines careful analysis of the interplay between technology design and socio-political processes with a wealth of practical experience to identify key challenges that efforts around IT for Good will always have to face.

– *Andy Dearden: Professor (Emeritus) Interactive Systems Design, Sheffield Hallam University*

Given the enormous influence and control of technologies over our lives, an ethical enquiry into their development, use and ownership is of vital importance. This book provides an incisive account of how state and market-led technologies have exacerbated socio-economic and environmental injustice, and conversely, how technologies based on the ethics of plurality, diversity, power-based equality, freedom and participation can help the movement towards justice and sustainability. Seth's call is not for rejecting technology, but for paradigm shifts towards more socially engaged technology and technologists.

– *Ashish Kothari: Kalpavriksh, Vikalp Sangam and Global Tapestry of Alternatives*

Professor Aaditeshwar Seth has spent years developing technologies through Gram Vaani, a social enterprise delivering a voice-based social media platform in northern India. Based on wide-ranging scholarship and hard-won experience, he counters market values with an approach to social impact that takes ethics and socio-technical theories seriously. If you're a technologist hoping to contribute to social good, this book will keep you honest!

– *Kentaro Toyama: Professor, School of Information, University of Michigan*

What comes out most importantly in the text is Aadi's two-fold firm conviction – one, that a technological community committed towards social good is indeed possible; and two, that dividing lines across technologists and ordinary people can be bridged, and this is what he has argued for. I hope that the technological community engages with these arguments.

– *Rahul Varman: Professor, Department of Industrial & Management Engineering, Indian Institute of Technology Kanpur*

For Gram Vaani, Stuti, and Iram.

Contents

About the Author

Aaditeshwar Seth is an Associate Professor in the Department of Computer Science and Engineering at the Indian Institute of Technology Delhi, and co-founder of the social technology enterprise Gram Vaani. He is passionate about building appropriate technologies and participatory tools that can empower marginalized and oppressed communities to collectivize and voice themselves. Several million people have directly touched these technology platforms. Over 150 development organizations worldwide have drawn upon the work done by Aaditeshwar's team at Gram Vaani and his students from the ACT4D (Appropriate Computing Technologies for Development) research group at IIT Delhi. Many elements of their work have also been adopted by government departments and have influenced the use of technologies for development in the social sector. He is a recipient of the ACM SIGCHI Social Impact Award for 2022.

Foreword

The use of digital technologies has transformed much in the world over the last three decades. However, has it made the world better? Has it reduced or increased inequalities? Have the world's poorest and most marginalized really benefitted? This wide-ranging and fascinating book seeks to address these and other crucial questions about the role of digital technologies in society, and aims to suggest ways through which positive changes can be implemented to make the uses of these technologies fairer and more equal.

The book is rare in the ways through which it crosses boundaries: written by a computer scientist it explores the relevance of social and political-economic theory; crafted by an Indian, it draws heavily on European literature. Much of the theoretical framing is thus situated within Aaditeshwar Seth's explorations of the works of European authors such as Marx, Foucault, Castells, Latour, and Gramsci; his basic demand for a paradigm shift in thinking about these issues likewise draws heavily on the USAn Kuhn's notion of scientific revolutions. His book is also enriched through a combination of this conceptual research with the experiential evidence drawn from his own practical engagement on the ground.

The potential agents of change for Seth are the technologists themselves. These are for him the engineers, designers, researchers, and managers involved in the digital technology sector. The book is intended to provide them all with a framework that can help change their practices – if only they will listen. He takes the reader on a journey that begins with understanding the importance of social goods underlain by an ethical foundation embedded within the traditions of humanism. The book then explores why traditional design processes have their limitations, and the need to change existing power structures so that they can instead be shaped to create more equal societies. He hopes that no technologist would want their labour to lead to harmful outcomes, and thus explores the structures and ideologies that limit their potential to design and implement projects that can be considered to be socially good.

This is not just a reflective and interpretative framework; unlike so many other recent academic works in the field it also has a profoundly normative stance. It suggests what should be rather than just what is. Indeed, the word 'should' is mentioned 178 times in the book's 220 pages! In a challenging account of ways through which disempowering 'paradigms' can be overcome, he suggests 17 questions that all technologists could think about if they really do wish to bring about 'good' change (pp. 158–159). These might usefully be stuck on office walls (in the real world) or embedded in software and posted on social media used by technologists (in the virtual world) to serve as reminders of the role that they as individuals can indeed play in making the world a better place.

Throughout, the book draws on specific examples and case studies, mainly drawn from Seth's own experiences. At various points in the book, he thus highlights the many challenges associated with the introduction of the Aadhaar unique identity system in India. He also draws extensively on the work that he has led in developing Mobile Vaani (supported mainly through Gram Vaani of which he is a Co-Founder), which is a federated network of voice-based participatory media platforms intended for less-literate users to share and discuss common concerns and solutions with each other.

In short, this book deserves to be widely read. It combines the author's passion and enthusiasm for the potentials of digital tech to be used wisely to help create better and more equal societies, with his understanding and realization of the many challenges that have to date prevented this from happening. Technologists across the world, and especially in India, are well placed to learn from and work with him, to begin to craft that better society.

Tim Unwin
26 January 2022

Tim Unwin CMG is Emeritus Professor of Geography and the founding Chairholder (since 2007) of the UNESCO Chair in ICT4D at Royal Holloway, University of London. He was formerly Secretary General of the Commonwealth Telecommunications Organisation (2011-2015) and Chair of the Commonwealth Scholarship Commission (2009-2014). His books include *Reclaiming Information and Communication Technologies for Development* (OUP, 2017) and his edited *Information and Communication Technology for Development* (CUP, 2009).

Preface

This book has emerged from confronting what appears to be a prevailing absurdity in the world today. We are surrounded by social problems of poverty, inequality, the environment, and many others, yet technology is scarcely deployed to directly address these problems. Technologists are more excited with getting advertisement predictions correct, or creating more addictive technologies, or improving technology infrastructures with little reflection on the uses to which the infrastructures are applied, and they assume that somehow magically these innovations will make the world a better place. Many of these innovations may however be entirely unnecessary, or may even harm users and society in general. Yet the world seems to be caught in a paradigm paralysis of continuous technology innovation without a moral compass to define worthwhile purposes of the innovations. A marginal category has indeed emerged of technologies for *social good*, but this space has remained small so far even though social good should have been the primary goal of technologies from the start. Even within this marginal category, although a growing brigade of technologists seem to be stepping in to address various prevailing social problems, they often get it wrong and create technologies that disempower the people they were meant to support. Yet the persistence of many such do-gooders remains unshaken. Voices and systems that would be truly empowering for people are sidelined in the presence of an orchestrated hype of buzzwords such as digitization, artificial intelligence, Internet of Things, smart cities, digital financial inclusion, etc., and their associated technologies are often deployed without contextual considerations which invariably worsens the entrenched structures of inequality.

What explains this absurdity of the world, of society, of technologists? How do well-meaning technologists end up building systems that harm people? Why does it always seem like an uphill battle to do what clearly seems to be the right thing? What should change so that genuine social good which avoids and prevents exploitation and disempowerment becomes the unanimous goal for technologists and society to pursue? I have been trying to walk this path for nearly a decade and a half of using technology to address social problems, and these questions have come up time and again. They come up in teaching where I feel we are failing to nurture a desire among students to use their skills for the wider benefit of society and to critically question the impact that their work has on society. They come up with our work at Gram Vaani, which has been incredibly difficult to scale in the presence of hype and problems created by other technologies, and has also humbled us with the complexity of bringing positive change in the lives of people. The questions also come up in professional research circles where research communities such as Information and Communication Technologies for Development (ICTD) that have drawn attention to problems created by technology have continued to remain small, while mainstream technology research continues vigorously to innovate systems with little interest in governing how these innovations get used. In my attempt to find answers to these questions in books, papers, and from my own experiences, I would not say that I have been especially successful, but I do feel much better situated now to understand the absurd ways of the world and I am convinced of two things.

First, overcoming these absurdities requires a paradigm[1] shift in how technologies are designed and managed. Technology design and management should be done with social good as the primary goal. The current paradigm of innovation which is driven by markets instead of morals, focussed on narrow values such as cost and time efficiency, is inadequate to solve important social problems. Second, to bring about a paradigm shift towards thinking of technology as a tool for social good requires the technologists to change themselves, that is, embrace a change from within – in their ethos and ways of working and thinking, rather than being guided by external regulations or value-less institutions like markets. This is why this book is addressed to technologists – the people who design, build, manage, and research technologies – to understand the current paradigm where technology often disempowers the weak, discover new rules and methods that an alternate paradigm of empowerment and equality should adopt, and build a strategy to bring about this paradigm shift. My hope is that these thoughts can be useful for technologists who, like me, may be feeling just as perplexed in seeing their labour leading to outcomes that at heart feel wrong, and join hands in charting a road where technologies are used appropriately and unanimously for social good.

This book is not a criticism of technology. Technology has indeed led to significant progress in building a healthier, empathetic, and more connected world. My attempt, however, is to understand what factors shape the outcomes that arise from technology, and how can they be controlled by technologists and society, so that disempowering effects can be avoided. It is not a recipe through which technologists can always align their work with social good, but some ideas and pathways outlined here may help us together discover better ways to move forward.

[1]I use the word *paradigm* in the same sense as Thomas Kuhn introduced in *The Structure of Scientific Revolutions*, as the dominant techniques, values, rules, and theories, which identify a particular framework in which science or technology is developed.

Acknowledgements

This is a book about values of technologists, and it would not have been possible without the people surrounding me who have shaped and informed these values for technologists like myself.

I want to thank the incredible Gram Vaani team for nurturing a space that allowed us to learn, make mistakes, and emerge stronger with the bold vision of empowering marginalized groups through technology. In no particular order, this is all due to the inexhaustible energy of Vijay Sai Pratap for taking over the reins of Gram Vaani which gave me space to reflect on our work and to put this book together; Sayonee Chatterjee and Sultan Ahmed who have always reminded us of the ground realities to spot gaps that might exist in our work and to overcome them; Kapil Dadheech and Rachit Pandey for leading the development and maintenance of the technology infrastructure that powers our work; Rohit Singh for patiently identifying synergies of our work one-by-one with literally hundreds of partners; Paramita Panjal for being our team's internal moral force reminding us to first be empathetic with one another before we can build more empathy in the world; Rohan Katepallewar for continuously finding new applications of our work and spawning exciting novel directions; Dibyendu Talukder and Praveen Kumar for setting high standards for our technology capabilities along with team members Ankit Kumar, Sohan Madhana, Aman Verma, and Prince; Rohit Jain, Vinod Maurya, Sujeet Kumar Choubey, Vishnu S, and Shiv Prakash Maurya for making sure that our technology services keep chugging along; Deepak Kumar and Rajeshwari Tripathi for patiently working through one of our most complex projects in Bihar; Brejesh Dua, Prashant Choubey, Matiur Rahman, Subodh Patra, and Ramjan Ali for maintaining the quality of our projects; Lamuel Enoch, Bruno Richardson, and Eswaramoorthy for independently managing our work in Tamil Nadu; Sangeeta Saini and Saraswati Kumari for being the strongest and longest standing pillars of our content team, committed towards upholding the quality of our work along with their team members Vasanti Kumari, Ritu Singh, Preety Kumari, Madhubala Pandey, Shweta Sharma, Suresh Kumar, Dinesh Rautela, Akash Anand, Mohona Dasgupta, Sunidhi Raj, Sonali Samal, Anjali Kumari, Shilpee Minz, Deepak Jaiswal, Anand Kumar, Anika Parween, Aman Gope, Akhilesh Kumar, Naweed Ali, Nasia Raunaque, Sumitra Kumari, Rohit Paswan, Rishikant Pandey, Aman Anurag, Rajeev Ranjan, and Juhi Mishra; Amrita Ojha, Ashok Sharma, Rafi Ahmad Siddiqi, Santosh Kumar, Amarjeet Kumar, Deepak Kumar, Zulfaquar Ali, Deoraj Pankaj, Mehtab Alam, and Sanjay Kumar for being our eyes and ears on the ground and our faces to the community; Akshay Gupta and Esha Kalra for providing strong research and outreach support; Veer Singh, Kanika Wadehra, Asha Gowda, and Govind Bisht for making sure that none of us are inconvenienced in our work and stay together as a team; newer team members including Abhideep Singh for rapidly imbibing the values in our work and pushing it further; erstwhile team members who strengthened our foundations, particularly Vani Viswanathan, Dinesh Kapoor, Lokesh Kumar, Kamesh Babu, Rohit Sharma, Shoaib Rahman, Ritesh Datta, Vidya Venkat, Orlanda Ruthven, and Roshan Nair; and most significantly our founding team who put together the vision and charted a path which the rest of us are still following – Aparna Moitra, Balachandran C., Mayank Shivam, Parminder Singh, and Zahir Koradia.

The Gram Vaani team extends to its hundreds of volunteers on the ground who proved their mettle especially during the COVID-19 pandemic by supporting community members to get access to food, cash, medical attention, transportation, social entitlements, and basic human dignity. It is through our volunteers that Mobile Vaani has upheld its standards of equality, humility, good journalism, and service.

All of Gram Vaani's work of course would not have been possible without support from numerous donors, investors, and partners, and the trust they placed in us. The list is really too long for me to mention all the people who in small and big ways have contributed towards it, but I specifically want to mention a few names who have been instrumental in shaping our work and vision: Anuragini Nagar, Arti Jaiman, Arvind Singhal, Ashok Shukla, C.S. Sharma, Daphne Luong, Harlan Mandel, Helen Hua Wang, the Indian Angel Network, Jean Drèze, Jessica Mayberry, the Knight Foundation, Lisa Braden-Harder, Mamta Kohli, the Media Development Investment Fund, Mira Johri, Nivedita Narain, Poonam Muttreja, Rajendran Narayanan, Rajiv Khandelwal, Rakshita Swamy, Reetika Khera, Rupsa Malik, Sajan Veniyoor, Sasa Vucinic, Sashwati Banerjee, Soham Mazumdar, Suhel Bidani, and Syed Karim.

I next want to thank my students who indeed are the ones to have uncovered the insights that I have simply threaded together into this book: My PhD students Zahir Koradia, Aparna Moitra, and Dipanjan Chakraborty who made their research and Gram Vaani's practice one and the same; Anirban Sen and Amit Ruhela who brought forth these ideas to new domains; research associates Aman Khullar, Aravindh Raman, Ashish Sharma, and Piyush Agarwal who helped deploy and evaluate many new innovations; and the huge army of undergraduate and master's students who have over the years provided strong support to pushing our research forward. I am sure all my current students will also follow in these footsteps and set new benchmarks for research that directly contributes towards social good.

Inspiration from role models is what keeps us going. Over the years, the ICTD community became my home and provided no shortage of motivation. In particular, I want to thank Bill Thies, Richard Anderson, Neil Patel, and Rikin Gandhi for their inspirational innovations and persistence; Lakshminarayanan Subramanian, Kentaro Toyama, Neha Kumar, and Melissa Ho for their commitment to holding the community together; Revi Sterling for pushing these values into broader agendas; and so many other colleagues for making ICTs for development into a genuine force for good.

Finally I want to thank all my colleagues from IIT Delhi at the Department of Computer Science and Engineering and at the School of Information Technology, for providing space to combine research with practice and never discouraging me from stepping outside of the ivory tower. In particular, I want to thank Huzur Saran, Sanjiva Prasad, and Subhashis Banerjee for their support and mentorship in this journey.

Coming to the book specifically, I am especially grateful to my PhD advisor Srinivasan Keshav for patiently reading through the entire manuscript and providing extremely detailed comments. Keshav is truly an inspiration and a life-long mentor for me – I give all credit to his tutelage for teaching me how to think, shaping my PhD research, and inspiring all his students to conduct high-quality and high-impact research. I can only hope to provide similar guidance to my own students. I am also extremely thankful to Balaji Parathasarathy, Nandana Sengupta, Amit Nanavati, long-time collaborator Mira Johri, Aditya Prakash Rao, and my students Saurabh Jain and Amit Ruhela for their extremely thoughtful feedback and suggestions which strengthened the book; Andy Dearden for his practical suggestions to strengthen my arguments and to link them with many other writings; and Pratyush Chandra for first pointing me towards literature on cybernetics and its connection with worker movements, which helped me connect the dots across these seemingly disparate topics.

I want to sincerely thank Tim Unwin for contributing a very kind and thoughtful Foreword, and to the Emerald Publishing team led by Kimberly Chadwick for all their attention to detail and taking the project forward.

My family deserves the most credit for supporting me in completing this book and for persevering in my work. Whenever I have felt defeated, I have not had to look further for inspiration than the strength of my mother Vasundhara Seth and the righteousness of my late grandfather Ramesh Chandra Seth. Whenever I have grappled with understanding new concepts, my late grandmother and teacher Vimla Seth has always shown the light. My wife Stuti Khanna has not only contributed directly by being the one to suggest that I write this book, even proposing its name, and for pointing me towards many valuable references, she has also kept me grounded and realistic all these years, and has patiently tolerated all my craziness. My daughter Iram Seth-Khanna is the most special person in my life – an activist to the core with a fierce sense of justice and always asking tough questions. This book is for her.

Chapter 1

Introduction

Computer technologists may appear to rule the world today. Technology companies have among the highest market valuations. Students aspire to turn into entrepreneurs and build the next tech unicorn. No industry functions without information technology. Informatization of production processes has, across industries, disrupted the very nature of work through increased automation and improved precision. Even the shift of the public sphere of discourse to privately managed online communication platforms has made democratic processes subservient to the design and management policies of these platforms. Large technology companies are so powerful today that they are able to resist pressures by governments to bring them under control, and threaten to withdraw instead of complying with regulations (Clayton, 2021). Investors alike are giving way to the founders and managers of technology companies, and allow themselves to be steered in whichever direction the founders want as long as the investors get good returns (Acemoglu, 2020; Surowiecki, 2012).

Yet, there is an increasing alienation among technologists. White-collar workers in technology companies face the same types of alienation as blue-collar workers face in factories (Healy, 2020). They have little say in the specifications of the software they write, or the final output that is created from the atomized technological components which they build. Even workers involved in creative processes with building or managing cutting-edge technologies face an existential dilemma of having no control over how technologies produced by their labour end up getting used by people, or being put to undesirable uses by their companies. In fact, in the name of making economic processes more efficient or increasing the safety and security for society, technologies are increasingly being used as instruments of surveillance to control populations, curtail freedoms, and shape consumer behaviour. This disempowers the very people whom the technologies were meant to support. The conflict is clearly visible with employee protests at companies such as Facebook and Google (Issac, 2019; Shane & Wakabayashi, 2018; Thompson & Vogelstein, 2018), the formation of white-collar unions in the Silicon Valley (Conger, 2021), and a backlash against the so-called big tech in many quarters of both society and the state (Doward, 2018). There is so much skepticism about

Technology and (Dis)Empowerment: A Call to Technologists, 1–7

doi:10.1108/978-1-80382-393-520221001

these technologies that many technologists who design and manage them advocate against their own families using them (Kulwin, 2018; Wong, 2017).

Even in scholarly academic spaces which are valued for the freedom and creativity they allow, there is increasing alienation arising from indirect control exercised by the dominant capitalist system over research direction (Healy, 2020). With pressures of getting research grants, demonstrating technology transfer to the industry, and metricized performance assessments in terms of publications and citations, many researchers are unable to change research directions or influence the ways in which their work gets used towards more meaningful goals for society. Rather than being able to solve social problems, researchers are realizing that optimizing the use of technology towards purely economic goals is clearly inadequate, but are able to do little about this disconnect.

Speaking from my own experience as an academic, and that of many colleagues, an alienation is arising even in teaching. Decisions by our students on what courses to credit, how much effort to put, and a long-term commitment towards projects, all seem to be driven by an economic rationality which results in alienating us from teaching as a means to equip students to solve social problems (Mehra, 2021). What could be a genuine form of exchange of knowledge and ideas between students and teachers becomes a mere transactional exchange of know-how and information that is devoid of any social connection between us and them.

Braverman documented long ago that white-collar work done sitting in offices may not be very different from blue-collar work done in sweatshops or on assembly lines (Braverman, 1974). Even the creativity allowed among elite technologists, or the high wages commanded by them which enables them to voluntarily choose what to do with their time, may not be sufficiently satisfying if they are unable to connect their work with the rest of society. Marx's concept of humanism explains this alienation successfully, with its axiomatic basis of humans as workers who produce for society, and derive their humanism through positive social relationships fostered by the production process and the produced output (K. Marx, 1844). These social connections created between workers when they cooperate with one another, and between the workers and those who find genuine *use-value*[1] in the produced goods and services, is what make them feel more human. Alienation arises if social or economic systems modulate these relationships and force them to take restricted form through domination or instrumental or exploitative use of others. Alienation is unsustainable since it makes us less human. Humanism implies the re-establishment of positive social relationships in their full form, through production that meets genuine use-values of society, through production processes that are not coercive, and that connect producers and consumers with one another in multi-dimensional social relationships of mutual understanding and cooperation instead of straightforward uni-dimensional economic

[1]I refer to genuine use-value as arising from a commodity when it fulfils a societal need, its use or production does not harm anybody, and these processes are also unmediated by instrumental mechanisms such as advertising.

relationships. Anti-humanist systems that harm people by creating negative social relationships alienate humans from humans, and are unsustainable.

Technologists are no different. Higher wages, sanitized white-collar office-spaces, flexibility of work, and scope for creativity, may give momentary illusions of autonomy and freedom to technologists, but their fundamental human nature will surface sooner or later to question the inevitable alienation they would experience if they continue to distance themselves from society. This belief forms the basis of this book, meant to provide pointers to technologists to re-acquire a humanism for themselves that is getting lost, and further to build and deploy technologies that can increase humanism in society as well. I equate social good with such a humanism and outline some steps that technologists can take to do social good.

In Chapter 2, I describe the context of current social and economic systems. The hegemony of capitalism has entrenched an ideology of self-interest and individualization, which reduces humans to economic agents rather than beings that draw their humanity from creating social relationships that benefit others in society. In coalition with the state and the apparatuses of media and digital capitalism that shape thoughts and behaviour, even undeniably negative values of exploitation, lies, and deceit manage to remain hidden from plain sight. Likewise, authoritarian socialist regimes that draw their ideology from equality and cooperation among all members of society, are no different in viewing people as mere cogs in a machine, with the use of propaganda to conceal the underlying contradictions. Technology, built and managed by technologists, has thus been reduced to an instrument to sustain these anti-humanist social and economic systems. Technologists, too, are encouraged and trained towards building systems that produce efficient economic relationships or tools that bring order in societies through control, rather than work on technologies than can foster cultures of humanism. Technologies produced by them therefore only reinforce the dominant ideologies.

Yet technology is also very powerful. Digital communication platforms have bridged physical distances between people. Information platforms have made knowledge more easily accessible to people. Transactional platforms have facilitated meaningful social relationships between producers and consumers. However, when such technologies are entirely driven by an economic rationality rather than humanist principles then they can have disempowering effects. Digital communication platforms such as Facebook have not done enough to prevent their appropriation by hegemonic divisive powers which increase social distancing in society.

The forms of knowledge commonly communicated over most information platforms on the Internet have served to entrench dominant ideologies of economic rationality in society. Transactional platforms such as Uber have sided with supporting relationships that are exploitative of the producers since their own economic sustainability is contingent upon surplus value that emerges from this exploitation. Technology is clearly a double-edged sword.

Can technologists ensure that technology does not merely provide tools to sustain the entrenched structures of today, and rather restores humanism to society? I believe that technologists who build and manage these technologies, are in an increasingly powerful position to bring positive changes for two reasons. One, they can increasingly control how their technologies operate so that the social and economic relationships, and knowledge, intermediated by their technologies, create more cooperation and pluralism in society, and less exploitation and domination of the weak. Two, as they gain increasing importance in social and economic spheres, they can dictate what technologies are actually built in the first place, to shift technology production to forms and spaces that restore humanism.

Chapter 3 therefore raises the question of what is good for society – what should technologists aim for, to bring more humanism in the world, and through this process to also build a new utopia of work for themselves? To answer this, I suggest reducing the *social good as humanism* argument to an ethics-based foundation. This can help debate in a principled manner about which values define social good, how to handle competing values, adherence to the values in both the means through which a technology is designed and the ends to which it is deployed in society, and understand the (dis)empowerment effects of technology about who benefits or who is harmed by it. Ethics-based reasoning can thus equip technologists to answer questions about what technologies to build, and how to minimize harm that may arise from it. Chapters 4–6 build upon this in more detail.

Chapter 4 describes an ethics-based framework to diagnose whether a particular technology is internally consistent in the values that govern its goals, design elements, and management practices. I discuss several aspects of technology design and management where ethical questions frequently arise, such as persuasive user interfaces that nudge user behaviour towards predictable directions, algorithms that encode objectives and quantify categories in a formal manner that can be discriminatory and unfair, data-driven algorithms that may reproduce societal biases, and consequently shape social relationships intentionally or unintentionally in ways that may disempower the weak and reduce humanism. I discuss the ethical values that govern these design and management aspects, identify some values that should straightaway be rejected as anti-humanist, and whether the rest are internally consistent with one another in a given technology system.

Chapter 5 examines the interface between technology design and socio-technical management in more detail, based on the observation that even a robust ethics-based and value-centric design of technology may not be sufficient to prevent harms arising from it. The complexity of the world and unpredictable uses arising from the affordances and malleability of most technologies make it imperative that a tight feedback loop is maintained to track how a technology gets used in practice, and this feedback is then used to manage the technology more effectively by constraining its harmful uses. I discuss this through my own experience over the years with being closely involved in guiding the development of Gram Vaani's platforms for participatory media. New challenges arose, and still continue to arise, at the socio-technical interface when these platforms are used and appropriated by people seeking to meet their goals.

While Chapters 4 and 5 are more focussed on avoiding harm arising from technology, Chapter 6 goes back to the question of what technologies to build in the first place – are there any core values that the design and management processes of technologies should espouse to restore humanism and do social good? I draw a relation between anti-humanism experienced in capitalist regimes as arising from an accumulation of power. Technologies of governance used by the state similarly rely on the monopoly of the state to use violence as a means of exercising power over people to control them. Social good as humanism will always be in-congruent with relationships of domination or those that disempower the weak. It is therefore necessary to build technologies that transform social relationships to lead to the distribution of power towards power-based equality in society. Building upon the wide literature of power, I discuss commonly used techniques that lead to the accumulation of different types of power, and collective action as a means to counter such centralizations of power. I suggest that a rationality based on the economics of power can be more successful in discovering pathways of doing social good than trying to find solutions through economic markets or following directions laid down by authoritative socialist structures. Transforming the entrenched power structures in society is however a political project. Even if technologists move away from their general apolitical attitude to taking a conscious political stand, many challenges are likely to confront them in walking down this path. Chapters 7–9 explore this question.

Chapter 7 outlines the organizational, economic, political, and societal structures within which technologists operate, which constrain their ability to uphold values of humanism in their work. Functionally segregated workspaces built upon the principle of the division of labour prevent technologists from establishing a direct connection with users and various stakeholders who interact with their technologies. This growing social distance between technologists and the end-users contributes to a lack of understanding by technologists of the true impact of their technologies. Economic pressures to provide increasing returns on capital and to counter competition, prevent technologists from testing their work adequately to ensure safe and productive use of their technologies for society. Carefully crafted ideological narratives by both the state as well as the corporate sector about the transformative potential of technologies to benefit society makes it difficult for technologists to see through this simplistic view of technology optimism. They fall for this mistaken belief and labour towards producing seemingly useful technologies but which in fact may be harming society more than benefitting it. Regulatory capture or simply hiding undesirable-outcomes arising from technologies further prevents democratic governance and economic markets to control what technologies get built and how are they used. Societal segregation with growing inequality, in which white-collar technologists predominantly form a part of the elite, also contributes to a disappearance of social relationships between technologists and users.

Chapter 8 presents several mechanisms to counter the problems presented in the previous chapter. Communication platforms to connect technologists with the users of their technologies, collective action by technologists to gain a say in the management decisions of their corporations, cooperative forms of ownership where management control rests in the hands of the workers, exercising lawful control over intellectual property to prevent its use for oppression of the weak, and financing mechanisms that can enable new initiatives grounded in humanist values to successfully compete with the dominant hegemony of capitalism, are some emerging exciting developments underway in different parts of the world. None of this is easy, though. Not only are such counter actions difficult to initiate and sustain by technologists for the reasons listed in Chapter 7, bringing about a transformation of work for technologists requires the participation and support from society itself.

Chapter 9 discusses steps for such a transformation in partnership with wider society. Human societies have already developed sophisticated institutional mechanisms to govern themselves in accordance with the values in which they believe. If a society embraces new values, then these can make their way to markets when consumers begin to demand these values from the companies with which they transact. Markets have no inbuilt ethics of their own, but are agile enough to include new values in capitalism. Alternately, citizens in democratic states can demand the government to enforce these values through regulatory means and in turn make markets adhere to these values. Or societies could build new systems of governance and the economy to replace capitalism. The crux, therefore, may lie in building initiatives that can bring about a transformation in society in terms of the values it respects. Technologists may have a role to play here, in enabling participatory democratic discussions on values among society members, while respecting pluralism and diversity among the members. In this regard, I finally discuss the need for information and communication platforms that can promote such discussions for deliberative democracy, as well as help technologists remain better connected with the users and other stakeholders interacting with their technologies. I highlight gaps in the current systems of participatory media, and what characteristics information platforms should adopt to serve as a basis for bringing about an understanding among diverse social groups. These discussions can help define values that society democratically agrees to be good and useful for itself, which can further inform the ethics-based foundations of work done by technologists for social good. Technologists will then not be alone, their struggle to prevent their own alienation will join hands with a struggle by society itself to restore humanism and impose its social control over technology.

I have written this book from the perspective I have gained as a technologist myself. I see my own experiences as being reflected in the theories of people including Marx, Foucault, Gramsci, Habermas, and Amartya Sen, which have been connected with current society by people such as David Harvey, Cristian Fuchs, Partha Chatterjee, Andre Gorz, and Mike Cooley, and echoes with literature in ethics and technology that is closer to computer scientists, written by people like Norbert Wiener, Luciano Floridi, and Tim Unwin, drawing validation from the harms of technology brought to light by colleagues including Kentaro Toyama, and inspirational figures of activists like Jean Drèze, who remind us to remain grounded and be in touch with the people whom we hope to impact. Ideas from thinkers such as Jean Drèze, Amartya Sen, Thomas Piketty, and Arturo Escobar, on the dynamics of poverty and development, and humanitarians like Harsh Mander who relentlessly try to repair the fabric of society that ties humanity together, have also either influenced my actions over the years or find a strong resonance with my own experiences, and have shaped the thoughts in this book. With such a diverse scholarship, and given my limited training in the social sciences itself, I have not attempted to adhere to the traditions of any one specific academic discipline. The book is therefore a mash-up of ideas, thoughts, principles, and validations, borrowed directly from eminent theorists such as the ones I have named above, or derived deductively in some cases, or in other cases obtained inductively through case-studies from my own work and immediate observations of other technologies. These have been useful to me to provide a basis for my own way as a technologist, and I feel are also increasingly relevant in today's world that is dominated by technologies yet has a long way to go in orienting its moral compass to positively impact society. Most of the examples I use in this book are drawn from India and my own work, while the theoretical concepts on which I base my analysis often originated in other parts of the world. My hope is that this mix of geographies will make the arguments more widely applicable, and especially draw attention to similarities and differences in technology related issues in the Global South and Global North.

As a teacher, I hope my students will find this useful and align themselves towards meaningful work for society. As an entrepreneur, I hope my colleagues from Gram Vaani will understand the tremendous value that emerges from their day to day work towards which they have dedicated themselves. I also hope that my academic friends and colleagues, those in the social sector, and investors and funding agencies who have worked with me, will be able to understand the true motivation behind the work by me, my students at IIT Delhi, and team members at Gram Vaani. Lastly, the book is meant for technologists – those I may not know, but for whom it is most important to recognize their unique place in today's society and the transformative potential they hold to truly make this world a better place through their work. Technologists working in the ICTs for development space may find discussions about my experiences with Gram Vaani to be helpful in understanding the complexities of realizing impact through technology. Those working in companies that do not specifically identify themselves with doing social good or working towards social development may benefit from the discussions on ethical evaluation of technologies and how technologies may empower or disempower people. Technologists who genuinely do want their work to benefit others may realize why they need to build a broader understanding about the day to day lives of their users and how technologies shape them, to then design and manage technologies in a responsible manner. I hope this book will help technologists to see through the exploitative structures dominant in today's world, understand why these structures persist, learn how to transform these entrenched structures by identifying and building technologies for social good, and engage politically to support and scale these technologies, to eventually displace the hegemonic paradigms of technology production that result in technologies which disempower people.

Chapter 2

Contemporary Problems

According to Marx, humans derive their humanism from social relationships that are forged by non-coercive production processes and through the uses to which the outputs of their labour are deployed, to meet genuine needs of society. Despite technologists holding a dominant position in society today that brings to them high wages, creative work, and collaborative working environments, there is, however, an increasing alienation among technologists in not having enough control over how the output produced by their labour impacts society. Humanism implies that this output should create genuine use-value, that is, fulfil a genuine need in society without harming others, and thereby forge positive social relationships between the producers and consumers. What are these real needs of society that technologists should aim to meet? This is the subject of this chapter.

2.1 Apathy

I grew up in an educated and well-to-do family in Lucknow. We did not have any money to waste but we were also never out of means to afford all essentials or even any reasonable hobbies. My only early acquaintance with poverty was through the son of our domestic worker when I was a child. Neeraj was a dear playmate and we shared many adventures, as equals, but we were not in fact equal in anything else. I went to one of the best private English medium schools in the city, an hour away from our home in the city outskirts, while he went to a free public-funded school in the neighbourhood, but only when he did not have to take care of two of his younger siblings. I on the other hand had no responsibilities at all. I went to school, studied, played, and mostly did what I liked, while Neeraj almost daily was beaten up by his parents for forgetting to do some or the other household chore. I lived in a large house with my mother and grandparents, while Neeraj, with his two siblings, and mother and father, lived in one small room behind the garage in our house. We of course had LPG for cooking, while they cooked in a small porch outside their room on a smoky stove that burnt dried cowdung blocks. They had a kerosene stove too, but kerosene had always been a problem to procure, the public distribution system being riddled with corruption

Technology and (Dis)Empowerment: A Call to Technologists, 9–32

doi:10.1108/978-1-80382-393-520221002

and leakages back then, and Neeraj's mother preferred to use cowdung which was more easily available due to the proximity of our house to a village nearby with lots of dairy cattle. Despite all the problems, Neeraj was always a happy and delightful playmate though. Whenever kerosene was available, we would use a bit of it to light up rags tied on to sticks and poke them into crannies from which swarms of mosquitoes would then emerge. Our playground was vast – the entire neighbourhood, where we climbed trees, ran, played hide and seek in half-built houses, pretended to be military commandos who crawled into enemy territory through tall unkept grass, and other exciting adventures.

Neeraj's mother worked at our home, mostly with house cleaning and doing the dishes, which brought the family some cash and a place to stay in the city. His father never held a steady job though – sometimes he was a security guard at a nearby school, on some days he would sell spicy *chaat* wheeled around on his bicycle, while at other times he would go back to their village and help with harvesting in family agricultural land, but of which we only heard with reference to disputes between him and his brothers. No wonder the family preferred staying in the city, even though it was not easy to get by. We helped them with loans which they sometimes repaid, but the relationship between our families was more than purely transactional. We were almost like one large family, drawing closeness from overhearing each other's internal domestic arguments, supporting one another, celebrating festivals together, and enjoying easy other's moments of happiness and providing comfort in moments of sadness and hardship. When my mother was bed-ridden for more than a year after a near-fatal road-accident, Neeraj's mother was instrumental in taking care of her and getting my mother back on her feet. Eventually however they moved back to their village. Several years later we got the sad news that Neeraj had passed away after a long bout with a kidney problem. Even when we were together, he was often sick and knowing the maltreatment he received from his parents, the news did not come as a surprise to any of us.

Many years later, I went on to start my job at IIT Delhi and co-founded Gram Vaani. We routinely interact with poor people now through our work. Almost every day, I spend some time listening to the voices of people who record messages on Mobile Vaani platforms, about their problems, their joys, dreams, festivals, songs, floods, droughts, and broken dreams. When I now think about Neeraj and his family, their story seems like that of any other poor family who's voices I hear, with similar problems and unmet aspirations. It makes me wonder about what has changed. It makes me think about the determinacy created by the contours of poverty and inequality into which people are born and the predictable trajectories along which they live their lives. However, what troubles me most is if I had not been associated with Gram Vaani then would I be thinking about this at all?

I feel most people do not think much about poverty or its outcomes. Perhaps in India it is to do with the intimacy we have with poverty, whether we come ourselves from poor families like Neeraj's who have experienced poverty directly, or from well-to-do families like mine who have lived close with the poor. We know poverty is there and we know what it means. Despite this intimate experience and knowledge of poverty, what I feel odd is the apathy that well-to-do families have towards poverty though. I think anybody will find it hard to argue that such an indifference does not exist.

My Facebook feed, for example, is comprised mostly of posts by my friends from high school and my undergraduate institute, and it is dominated by photos of their children or funny videos or congratulations about awards and birthdays, or with the latest on the COVID-19 vaccine which was the biggest news when I wrote this section of the book. I generally saw more posts about vaccine planning than about farmer protests that were taking place simultaneously at the Delhi border, against new farm bills which were rammed through parliament a few months earlier without any due consultation (S. Narayanan, 2020). A divergence from this pattern in my Facebook feed did happen during March and April 2020 when the Indian government announced the COVID-19 lockdown with a four-hour notice and millions of low-income migrant workers were stranded in cities, out of jobs, cash, food, and transport to go back to their villages (Adhikari et al., 2020; Ruthven, 2020). The visuals of these migrant workers walking back, cycling, facing police brutality, sleeping in parks, with little clarity about train transportation, did dominate the media and my Facebook feed, but more as a surprise when people learned that Indian cities were actually run by migrant workers (I. Roy et al., 2021). Soon however all that was forgotten and the conversations reverted to lockdown life, politics, film star suicides, and elections.

I have been teaching at IIT Delhi for more than 10 years now. We do get students from economically weaker backgrounds, but the general demography of students is from well-off families who were able to afford good quality education or coaching institutes (or both), for their children. Throughout my career at IIT, I can recall of no undergraduate or masters student whom I knew, to have joined a job in the social sector, away from mainstream information technology or financial trading companies or management consultancies.

While this is entirely anecdotal, my students did a study of the coverage of several economic policies in prominent English language newspapers which had been selected from across the ideological spectrum in India. Aspects about these policies that were relevant to the poor, formed a minuscule fraction of the coverage in all the newspapers (A. Sharma et al., 2020). The coverage was dominated by aspects more relevant to the middle classes, or companies, or statements by politicians who were simply using these controversial policies to score publicity points over one another. Data from Twitter about these same policies also largely echoed the mass media: Since the poor were not on Twitter themselves, there seemed to be little reason for anybody else to represent their interests on Twitter. We also analysed parliamentary question hour data of discussions undertaken by political representatives, and oddly enough we got similar results – questions of relevance to the poor were asked much lesser than questions pertaining to other stakeholders or questions about procedural matters to execute different policies (A. Sen, Ghatak et al., 2019). Only in highly politicized economic topics, like the demonetization event in India of 2016, did the poor really figure. Even political representatives who get most of their votes through promises to the poor did not seem to be representing their interests.

I do not think that this indifference of the well-to-do towards poverty or the poor was any different in the 1980s and 1990s when I was growing up. I have always idolized my grandfather for his selfless community service and so many other qualities. He relentlessly campaigned for municipal works in our locality

– improved drainage systems, better roads, reliable electricity, safety and street lighting, etc. As a true Gandhian, he would conduct non-violent strikes in government offices and force the officials to promptly attend to issues of our neighbourhood. However, I do not recall my grandfather ever talking about poverty. Or for that matter, even parents of any of my school friends, or other families in our neighbourhood, ever discussing or doing anything about poverty. Poverty was not invisible to us because we clearly recognized that it existed and we saw it everywhere, living side by side and co-existing with it. We were also not surprised by its outcomes because we understood how it worked and what it implied, but it seems that in general we chose to do nothing about it other than occasional acts of charity or small contributions made to CRY or UNICEF.

Why is there so much apathy towards the poor, among the well-to-do? Maybe Indians, true to our dominant Hindu roots, accept poverty as preordained fate from the *karma* of previous lives. Or the poor are believed to be responsible for their own shortcomings. Or the well-to-do have enough of their own problems than to worry about those of others, and consider that paying taxes is sufficient. Or perhaps they are realistic enough to not take up daunting problems like poverty. Or they face *compassion fatigue* from seeing poverty all around them. Or maybe they do not see poverty as a problem at all and rather as a functional requirement for a well-ordered society as what was organized by our ancestors according to the caste system, and continues in a similar manner with new types of social hierarchies created by capitalism.

Whatever be the reason, this apathy of the well-to-do towards the poor is a problem. It reveals a weakness of humanism as an unfelt desire to support better lives for others. It encourages the well-to-do to outsource their morality to institutions such as the welfare state, rather than take any moral responsibilities themselves towards the poor. It blinds them also from thinking rationally in how systems that reproduce poverty may actually be harmful even to themselves in an entirely economically rational sense. In short, apathy towards others can turn into a spiral of decline in humanism, morality, and rationality, all of which are important if we are to solve any of the grave problems facing us today, for society does face many important problems.

2.2 Social Systems

Poverty is certainly a big problem facing our world today. Depending upon what statistics are used, anywhere between 300 and 800 million people in India, and about 1.9 billion people in the world, can be categorized as poor. Poverty impacts many other social realizations that people are able to achieve, including education, health, livelihood, and a dignified life. Much has been written about poverty and I would be naïve to even attempt to describe it in a short section. The book *An Uncertain Glory: India and its Contradictions*, by Amartya Sen and Jean Drèze, is probably one of the best one-stop reference to poverty and its associated problems in India (Amartya Sen & Jean Drèze, 2013).

Poverty, providing education and health care to everybody, and dignified livelihoods, are however not the only problems. Global warming, pollution, exploitation of the weak, intolerance of plurality, and inequality that underpins many of these problems, threaten the very existence of human civilization, and can set back whatever progress the world has made in becoming a better place.

Again, excellent informational resources exist to describe these problems and I will not attempt to summarize them here, but the question I want to discuss is whether existing institutions of governance and the economy are robust enough to address these problems or not.

Even Karl Marx was impressed with the productivity gains unleashed by capitalism due to competition in free markets that capitalist policies espoused. Free-market systems as advocated by Adam Smith were assumed to allocate manufacturing capacity to service consumer preferences in the most cost-effective manner. The financialization of the world economy was meant to make this even easier by guiding credit allocation towards meaningful and appropriate production for the goods and services desired by consumers. Democratic countries where people voted for policies and held their politically elected representatives accountable to negotiate effectively for these policies, were meant to prod the state and markets further in the direction that people desired. No wonder then that Fukuyama called this the *end of history* – a perfect self-regulated system to run the world in service of the real needs of the people. Are these systems really perfect though, and can they be trusted to solve the grave problems that we face today?

Colin Crouch in his book, *The Strange Non-Death of Neo-liberalism*, describes why the markets are actually not as competitive as they are popularly made out to be (Crouch, 2011). The competitive markets during Marx's time were dominated by a large number of small firms, but today it is rather a few giant corporations in any industry that dominate the markets. This monopoly or oligopoly, instead of bringing efficiency to the production process, or providing goods and services that people actually need, relies on advertising and marketing to create a demand for whatever is produced by them. Large corporations further use their market power to manipulate prices both upstream and downstream, to expand their market share through price manipulations meant to squeeze out competitors. They also manipulate the state to use public funds to build positive externalities that are convenient for their expansion, yet lobby for non-interference of the state in imposing taxes or penalties for any negative externalities created by them. In a constant search for new markets, neoliberalism also pushes for the privatization of public services to expand its reach and influence. Amartya Sen explains that Adam Smith's conception of the *invisible hand* has in fact been blown out of proportion to justify markets as an end in themselves (Sen, 2000). Smith himself drew attention to the dangers of monopolies and government lobbying, and the importance of qualities of humanity, generosity, and public spirit as being most useful to people, but which the economic exchange market does not incorporate (Matson, 2020).

Even while neoliberal capitalism tries to reduce the interference of the state, David Graeber discusses in *The Utopia of Rules: On Technology, Stupidity, and the Secret Joys of Bureaucracy* of how capitalism also needs the state to enforce laws that enable markets to operate (Graeber, 2015). Neoliberal capitalism influences the state to develop laws, or bureaucracies, that only large corporations are able to navigate and thereby manage to hold down competitors, evade consumer rights, and run misleading advertisements with little fear of repercussions.

In the name of development as defined by nations of the Global North, institutions at the global level, like the WTO, IMF, and the World Bank, enforce similar bureaucratic structures internationally that countries of the Global South are forced to follow (Escobar, 1995). These however promote the growth of Western corporations at the cost of developing countries being able to build their own local corporations, or they facilitate routes for the foreign ownership of local corporations, essentially amounting to neo-colonialism. In both cases, these institutions help provision the underlying rules and regulations that benefit large corporations and investors to keep expanding.

Joseph Stiglitz in *The Price of Inequality* similarly discusses how the growing inequality at both the individual level and corporate level in the United States, helps wealthy individuals and companies to further strengthen their positions through policy lobbying and political connections (J. E. Stiglitz, 2012). In the Indian context, Atul Kohli tells this story in *Poverty Amid Plenty in the New India* about quid-pro-quo crony capitalism: India's largest corporations provide electoral funding to political parties in return for licencing of natural resources at dirt cheap prices (Kohli, 2012). My own students have studied the evolution of government–corporate interlocks over time during different election years in India, and identified hotspots of government departments and corporate industry sectors that tend to be tightly interlocked with one another (A. Sen et al., 2018). Similar observations were made decades ago in C. Wright Mills book *The Power Elite*, of the nexus between layers of political elite, government officials, corporate CEOs, and military leaders in the United States (C. W. Mills, 1956).

Clearly, the internal workings of capitalism in practice seem to be based on manipulation and deceit, rather than on classic economic theories of free markets. With such a poor ethical record, it is hard to place trust in the current form of capitalism to solve the problems mentioned earlier.

Not only that, Marx has argued that capitalism's very basis is centred on the exploitation of workers (K. Marx, 1867). Capital draws its returns from the surplus value it extracts through the labour process. We witnessed this most recently through worker voices shared on the Gram Vaani platforms in the aftermath of the COVID-19 lockdown in India. Migrant workers who were forced to return to their villages when jobs closed down in the cities, slowly began to come back to the cities after a few months due to a lack of employment options in their native places. However, they faced worse working conditions than prior to the pandemic (Seth, 2021b). Several states suspended labour laws related to health and safety norms and working hours, in a misguided move to revive the economy by spurring the resumption of production with fewer available workers. Taking advantage of the pandemic situation, the central government also passed new labour laws by avoiding due parliamentary consultation, which give employers more room to stay outside the purview of meeting necessary compliances for worker safety and other norms (Sood, 2020). With employers emboldened by such signals from the state, and partly due to a slowdown in their revenues because of a lower demand for several goods during the pandemic, we were flooded with reports by workers of an increased workload, longer working hours, no overtime payments, growing casualization of work, and even a resort to extortionist tactics by employers who began withholding the approval of social security claims of experienced workers to prevent their attrition (Seth, 2021b). The outbreak of violence at Wistron's iPhone manufacturing factory when several hundred workers went on a rampage, did not therefore come as a surprise at all (T. Johnson, 2020). Given a chance, capitalism clearly resorts to its underlying fundamentals of squeezing workers by reducing wages. This room, created with an improved bottomline, is used either to corner the profits and provide a higher return on capital, or to reduce prices to crowd out immediate competition and secure higher returns when monopolization emerges in the future. In other words, competition does not necessarily bring about an increased efficiency in the production process, rather it results in a race to the bottom on working conditions by reducing wages and exploiting workers. What remains invariant is the compulsion of capital to secure returns for itself by extracting the surplus value created in the labour process.

None of this of course happens without the assistance by the state. The state ultimately provides the bureaucratic apparatus to uphold contracts that ensure a return on capital even if it may allow contracts meant to secure humane working conditions for the workers to be routinely flouted. Trapped by the fetters of globalization, countries in the Global South are caught in competition with one another, and choose to relax their labour laws and weaken enforcement of working conditions to attract capital for continued investments and generation of jobs (Hensman, 2011). The ability of workers to collectivize is also systematically weakened by using the state to thwart collective action as what happened during the Maruti incidence in 2012 in India when the government declared illegal a tool-down by workers at Maruti in protest of low wages (Ness, 2016). Laws are now additionally being imposed that make it difficult to build unions, and by allowing a growing contractualization and casualization of the workforce to go unchecked makes collectivization even harder. Expenditure by the state on social protection through unemployment benefits or alternate state-sponsored employment schemes and on welfare policies is kept at a bare minimum so as to avoid large-scale civil society unrest yet have the civil society continue to provide a reserve army of labour to capital. This was obvious during the COVID-19 lockdown as rural employment and other welfare schemes came under significant stress due to increased demand (Seth et al., 2020). Unable to meet the needs of millions of people out of work, people had no other option but to return to work with even less bargaining power than before. An increase in both corporate taxation as well as direct income taxation of the wealthy, to fund stronger social welfare and social security for citizens, is further vehemently opposed by the well-organized elite. Such a system that is fundamentally exploitative and justifies inequality, yet resists the creation of equitable means for people to realize opportunities, is unlikely to help solve the kind of problems we listed earlier.

Through steady lobbying by the elite, the welfare state in most countries has degenerated into a regime of low wages, poorly financed welfare schemes, and growing inequality (J. E. Stiglitz, 2012). Mass consumption required by capitalism is thus unable to be serviced by the disposable income of workers alone. A new solution was developed in the West of complex financial instruments to finance household consumption through debt that was ultimately funded by the elite owners of money capital. David Harvey explains in *Seventeen Contradictions and the End of Capitalism* of how this new form of consumption works, by first providing it for meaningless consumption that does not fulfil any real needs of society, and then when people default, which they inevitably will, it effectively ends up transferring assets from them to the owners of money capital, increasing inequality even further (D. Harvey, 2014). This debt fuelled household consumption caused the 2008 financial crisis in the West. Undeterred by these dangers however, the same modus operandi is expanding to other countries. New markets for debt are being created among the poor in developing countries aided by increasing digitization and financial inclusion which is necessary to track capital, and to make access to credit as easy and frictionless as possible (Gabor & Brooks, 2016). Both domestic as well as foreign owners of money capital stand to benefit from any fallouts. To bring more and more commodities into ever expanding markets, assets like land are especially encouraged to be formalized so that the poor can capitalize them (Soto, 2001), but this is more likely to lead to *digital land grabbing* of transfer of land to the owners of capital (Benjamin et al., 2007; GRAIN, 2020). Hardt and Negri have termed this universal world order as *Empire*, which wants to be present everywhere, transact everything, and dissolve any national or organizational boundaries that may hinder exchange (Hardt & Negri, 2000). It proclaims freedom and inclusion as its ideology, even though it is based on an actual denial of liberation through a strategy of formalization – once something is formalized, *Empire* steps in to control it and profit from it.

As Thomas Piketty has exhaustively documented in *Capital in the Twenty-first Century*, mechanisms of worker exploitation, securing a return on capital including across national borders, state–capital nexus to shape laws and policies that benefit the elite, and new market creation through debt financed consumption and behavioural advertising, ultimately lead to an increase in inequality across the world (Piketty, 2014). Even in the case of market failure, the protection offered to banks by the state in the 2008 financial crisis suggests that crises themselves can become an opportunity for a further increase in inequality by moving wealth from ordinary people to the elite. Stiglitz similarly discusses the impact that the increasing development of labour-saving technologies, such as through artificial intelligence, will have on employment around the world (Korinek et al., 2021). He postulates that it will lead to the concentration of higher skills in Global North countries that are home to the winner-takes-all big-tech enterprises, while also increasing inequality within the Global South countries due to a rise in unemployment. Several policy measures are suggested, such as re-distributive policies of taxation, social protection, and universal basic income, and pre-distribution policies to steer technology innovation towards labour-using rather than labour-saving technologies, without which un-steered capitalism may push the world towards increased inequality.

People's faith in capitalism has certainly been dented in recent times (Edelman, 2020). It is however yet to be seen whether civil society will be able to mount sufficient pressure through electoral mechanisms in democratic countries to reduce inequality, or whether it will lead to a sudden anarchic end to capitalism, or will it produce a gradual shift in the values of capitalism. What is clear however, first, is that unregulated capital is fundamentally exploitative of workers and only interested in securing a return for itself, by hook or by crook. Second, the current form of neoliberal capitalism seems to be based on assumptions about competitive markets even though markets are imperfect, assumptions about the benefits of *laissez-faire* even though large corporations influence the state in crafting policies conducive for them, and assumptions about intelligent allocation of capital for a better world even though financial markets are inefficient at best and manipulative of the public at worst in allocating money capital. Capitalism does not deny that it has no inbuilt ethics or a moral compass – it is guided only by the ethics of consumers, or regulations imposed by the state which in turn are shaped by citizens. Yet through behavioural advertising it relentlessly tries to shape the perceived needs and aspirations of consumers, and it uses its hegemony to attempt to corrupt the state or to vehemently resist interference of the state in regulating the market.

Will people through democracy be able to change capitalism or replace it with some other system? Partha Chatterjee explains in *I am the People* of the link between inequality and electoral politics in India (Chatterjee, 2020). After the economic liberalization in 1991, as socialism weakened and welfare policies were increasingly adopted, the state emerged as the dominant agency to help the poor by distributing wealth. Politicians especially at the regional level were able to leverage this regime and build welfare schemes targeted towards specific social groups, or control access to schemes through retail corruption and clientelism, in return for votes. Most of these schemes were primarily populist, aimed only at short-term relief, rather than to address any structural problems. No actual substantial gains were achieved by the people since no long-term problems were actually solved, and people therefore became increasingly dissatisfied. At the same time, as neoliberalism became stronger, the ability of the state to provide benefits to the poor became increasingly fiscally constrained. Further, a move towards greater professionalism in welfare services delivery embraced by institutions such as the World Bank and IMF meant that control of welfare policies shifted to administrators, particularly controlled by the central government, and the populist strategies of regional politicians became increasingly irrelevant.

The culmination of all this disappointment in the politicians and the state unfortunately led to the rise of the *popular leader*, the vision of a strongman who can muscle the *system* into effective service delivery. Even though this new political era of a popular leader cannot be effective in addressing poverty since it is ultimately constrained by the chains of capital, it is very dangerous because the hate-politics simultaneously embraced by the right-wing in the name of religion, nationalism, race, conservative values, and other such markers, risks tearing apart the social fabric within India. This form of authoritarianism relies heavily on the use of media and propaganda to create a semblance of sanctity and robustness of national institutions, even though it appropriates them as instruments for its own sustenance. A weak society with compromised institutions and ruptured social fabrics means that democracy, the only vehicle that had any chances of representing the interests of its people in a face-off against capital, will lose even the little power that it currently has and give a free-hand to capital.

Giving a free-hand to capital could take the world anywhere. If a handful of capitalists come to own more and more of everything, and the nation-state is increasingly unable to represent the people, like in India, it may imply the following. One, whatever the elite deems to be useful for the world, or values that they decide, will be chosen as the projects executed by the rest of world. Two, repression of dissent and uprisings will become the sole control function of the state, if the state continues to exist at all. From across the board, Harvey, Crouch, Piketty, Chatterjee, and even The Economist (2020), are putting their faith in the civil society to uphold democracy, and through that to keep a check on capital.

None of this is to say that the current systems of the state and markets have been a complete failure. Books like *Factfulness: Ten Reasons We're Wrong About The World – And Why Things Are Better Than You Think* by Hans Rosling convey the advances in peace, health, living conditions, and other important aspects achieved by the systems of democratic governance and economic markets, often through technology innovations fostered by these systems (H. Rosling et al., 2018). The remarkable short duration within which the COVID-19 vaccine has been developed through a collaboration between companies and governments is a testimony to this. Significant advances are being made with environmental technology to counter global warming too. However, even the pursuit of these goals requires guardrails on capital which can be imposed in only two ways: through regulation by the state which in turn needs to be democratically regulated by the people, and through the shaping of consumer preferences which in turn needs a pluralist democracy in which people can learn from one another. Without these guardrails, capital may be agile enough to lurch from crisis to crisis and manage to survive through the use of tactics like relocating capital from one place to another, inventing new technologies, shaping demand, and other means, but it will not embrace ethical means and ends unless these have some exchange-value that can be traded in the markets. Any goals that it does take up will rely on continued extraction of surplus value from the labour process by exploiting workers. It may solve problems like poverty or global warming without dispossessing the poor only if the people can wield the power to control it, but it will not address these problems in non-exploitative ways on its own. Its inherent nature for concentration is more likely to push the scales towards greater inequality and control over others.

Environmental Technologies

Can technology solve global warming? Solar and wind farms have emerged as green alternatives to coal and gas power plants. Data-driven operations in smart cities and smart homes claim to bring greater power efficiency. With other green developments like electric cars, environmental technologies that have emerged through the current systems of the state and markets seem useful to address the impending challenges of global warming.

The argument about social systems I have presented above is not against technological innovation *per se*, or even against market or state-guided technological innovation, but that the creation and adoption of such innovations unless regulated appropriately comes at the cost of dispossessing the weak. The manufacturing of rechargeable batteries, for example, requires metals like cobalt and lithium, 60% of whose supply comes from the Democratic Republic of Congo where children are employed in unregulated mines (McKie, 2021). Similarly, the decarbonization agenda has shifted e-waste management of technologies such as smart meters to countries like Ghana which exposes not only the workers there but also their family members and children to pollution flows from the open burning of waste material (Sovacool et al., 2021). This injustice of causing a degradation of human rights and the environment in the Global South is, of course, not restricted to environmental technologies but has been a feature of capitalist development over the centuries (Gonzalez, 2015). Even recent calls for de-growth economics for a sustainable world are effectively not compatible to address the environmental injustice that has been historically perpetrated by developed countries (Rodriguez-Labajos et al., 2019). Unless this injustice is rectified and a new regulated regime of technology development and use is put into place, environmental technologies will only serve to amplify the inequalities further.

Mechanisms like auditing of supply chains to guard against the use of child labour or poor working conditions or environmental pollution, have been proposed, but are known to have not worked well in other industries such as garments manufacturing (Theuws & Overeem, 2014; Tiwari, 2021). Greater international cooperation, monitoring, law enforcement, and even shaping of consumer behaviour to demand accountability on these fronts from manufacturers, are essential requirements so that new technologies do not produce dispossession.

2.3 Society

Social systems undoubtedly emerge from society, but once these systems become dominant they begin to exercise their influence on society and end up shaping its values. One of the greatest fallouts from the hegemony of the capitalist ideology has been the rise of individual self-interest, the economization of society, and a diminished sense of responsibility by outsourcing personal morality to regulatory institutions (Margalit & Shayo, 2020).

Self-interested behaviour creates an ideology of competition which does not remain restricted only to firms, but percolates to the individual level when people compete with one another for jobs and opportunities. While this may sound like a reasonable meritocratic approach to discover the best candidates, Michael Sandel in *The Tyranny of Merit: What's Become of the Common Good* explains how it falls short in the same way as the concept of free markets can be unfair (M. J. Sandel, 2020). The discovery of merit and accruing returns to it, suffers from the same problem as markets because competition among the candidates can be imperfect, depending upon the opportunities they have had at earlier stages in their lives. Mechanisms are needed to bring opportunities for growth and mobility to everyone in a fair manner, which requires interventions that go beyond a simple embrace of competition on the basis of merit. The poor, for example, cannot be blamed for being poor because of their own shortcomings; constraints in the wider social systems have greater effects on what people are able to achieve (Kinney, 2021). Sandel further describes how the merit-based sorting apparatuses in place especially for admissions to educational institutions in United States end up creating segregated classes which erodes solidarity in society and reduces diversity through which people could have built an awareness of the lives of others.

Collective action thus becomes harder to justify and execute within a capitalist ideology. Not only is centrality attached to individualization instead of cooperation, the outcome of any effort undertaken by an individual also demands to be justified in economic terms of the value realized by each individual. Mancur Olson in *The Logic of Collective Action* shows that when cooperation needs to be justified in terms of the economic gains that it produces collectively for the members, costs are incurred by only a few while others free-ride, which leads to conflicts in the equal distribution of these gains (Olson, 1965). This is especially difficult in larger groups when observability of joint efforts put in by the members becomes hard. This negatively impacts solidarity in the group, which, of course, suits capitalism since it weakens collective action that could otherwise challenge it. The capitalist ideology therefore does not glorify cooperative forms of organization like the ones described by Elinor Ostrom in *Governing the Commons: The Evolution of Institutions for Collective Action* which through non-market mechanisms are able to achieve democratic consensus and aggregate good for their members (Ostrom, 1990).

The role model endorsed by capitalism is that of a bold risk-taking entrepreneur who competes on the basis of their merit, gains wealth as a result of their efforts, and drives the world forward. Entrepreneurs, as we know, are not always well-meaning and do not derive their gains through innovation alone. Their true merit often lies in their ability to exploit opportunities created because of market imperfections, with no specific regard for equality, justice, reason, and deliberation in the solutions they develop. This creates the *enterprise society*, as termed by Michel Foucault, where each individual begins to consider themselves as a provider of services, transacted through capital, with no underlying ethics to set the direction for the deployment of capital (Sandoval, 2019). Institutions of schools, media, markets, and the state, all endorse this model to the extent that this underlying ideological hegemony of capitalism becomes normalized and goes unchallenged, crowding out alternate systems for social organization.

A rise of individualization leads to social alienation. C. Wright Mills explains in *White Collar: The American Middle Class* that growing individualization makes people use others as instruments for their own gain (C. W. Mills, 1951). A pervasive salesmanship mentality for instance marks entrepreneurs and successful people who use their own personality to convince others, including their own colleagues and bosses, about whatever they may be trying to persuade for adoption or purchase. This ultimately leads to a growing distrust in society that alienates people from one another.

Notwithstanding these concerns, corporate-owned commercial media creates hype for free markets and entrepreneurism since it too, ultimately, benefits from echoing capitalist ideology. Alternate systems are discredited quickly by conjuring images of totalitarianism, loss of freedom and growth, and anarchy and chaos. This leads to an alienated society that favours individualization over collectivism, and economization over humanity. Such a society whose values are shaped by capitalism is unlikely to be able to stand up and keep a check on capital, to solve problems of poverty, inequality, and exploitation that the world faces today.

2.4 Technology

Technology develops within the given social structures at the time, and is shaped largely by the contours laid down by these structures. In the current capitalist social system, the priorities for new technology development and use are defined through needs articulated by the markets, which requires the existence of current or future paying customers, and leads to allocation of funding for research and development which is in line with these directions. Technologies that challenge this social structure either do not have paying customers, or are deliberately constrained in their adoption and growth through funding criteria defined by the organized elite, who benefit from these structures, and are in-charge of maintaining them, or these technologies are out-competed by alternative technologies which are more in-line with the existing social structures, or they are simply co-opted by taking them over and altering their vision to bring it in line with the hegemonic social structures.

There are plenty of examples to validate this. Graeber suggests that disproportionately large defence funding into surveillance technology is responsible for the rapid growth of information and communication technologies (Graeber, 2015). Braverman describes how Taylorism and monopoly capitalism increased the need for monitoring and control of the production processes, value accounting, financial management, and marketing, which now form the bulk of the use of computer systems in various industries (Braverman, 1974). Similarly the imperative of globalization to coordinate manufacturing and marketing activities across borders, with the ability to allocate capital internationally in agile ways, favoured increasing investments into the development of the Internet. Powerful visions of alternate ways to use the Internet for a stronger democratic world order lie marginalized with the appropriation of these technologies by advertisers and the same political gatekeepers who earlier manipulated print media and the mass media. Shoshana Zuboff describes these processes in her book *Surveillance Capitalism* of how the need for behaviourial data generated through participation of people on Internet platforms has led to an increasingly centralized infrastructure that is more amenable to data analysis than decentralized infrastructures, and facilitates behavioural control by both marketers as well as by governments and politicians (Zuboff, 2018). This business model of advertising and behavioural control that monetizes freely generated behavioural data of the users, has the same characteristic as regular capitalism, to appropriate the surplus value generated by the participation of users on the platforms. Christian Fuchs discusses in *Social Media: A Critical Introduction* this surplus value of the unpaid labour expended by users on these platforms when they create and share content with one another. The unaccounted surplus also appears as a positive externality which the platform providers are able to monetize by making their platforms more attractive to acquire additional users (C. Fuchs, 2013). Gig economy platforms similarly make use of the surplus of positive externalities created through network effects to bring about greater concentration in the market, and use this increased market power to extract more surplus value from workers by paying lower wages to them. Ultimately, thus, technology seems to be just a tool caught within the social and economic systems in which it exists, unable to disrupt them or carve out new structures, and only entrenches them further.

Harvey is his essay *The Fetish of Technology* further points out that technology innovation itself does not require capitalism, as is made out to be through the media hype about the incentives for private gain that capitalism creates to foster innovation (D. Harvey, 2003). There are plenty of examples to show that innovation does not necessarily need private gains. In fact, being able to contribute to society for public gains is often a stronger motivation. It is capitalism, rather, that needs constant technology innovation for the creation of new growth opportunities by replacing old technologies with new, or for controlling the labour process to make the extraction of surplus value easier and more cost effective. In this regard, Langdon Winner in his essay *Do Artifacts Have Politics* describes technologies deployed by capitalists that facilitate factory supervisors to monitor workers better, and to atomize the production process in ways that reduce the chances of socialization among workers to disrupt it (Winner, 1980). Yet another route through which capitalism benefits from technology was demonstrated by none other than Charles Babbage, the father of computers. In his book *On the Economy of Machines and Manufactures*, Babbage justified the lower costs that emerge from a division of labour, compiled on the basis of visits made to over 100 factories (Babbage, 1832). He showed that breaking up a production process into simpler independent tasks, which require differentially skilled labour, is cheaper than if the entire process were completed by a single person having higher skills. Replacing labour with capital intensive technology is still another way, which creates unemployment and reduces the bargaining power of labour to resist further exploitation and the continued expansion of capital.

Computers are in fact valuable tools in task atomization and are the ultimate bureaucrat for the modern assembly line due to both the precise control they can exercise for well-coordinated work by different workers, as well as the possibilities for automation that they create. The history of industrial automation shows how *informatization* of the production process to make it computer controlled led to a skills polarization by having skilled programmers on the one hand and unskilled machine operators on the other hand, to replace skilled machinists who manually operated the machines earlier. This wrested skills away from the machinists by encoding their knowledge into programmable machines, as well as control away from the workers by putting it in the hands of the management. Computers thus became tools of domination by the management over workers. This was the main reason for preferring computerized automation in production systems even though it was only after much trial and error spread over several decades that computerized systems were able to provide the same production quality and cost as alternative systems that allowed more worker control (Braverman, 1974; Nobel, 2011).

Technology effectively thus becomes a mirror of the prevalent social systems and amplifies the values that these systems embody themselves. For example, platforms like Facebook developed within the capitalist context, echo the ideologies of capitalism. Its highly metricized user interfaces emphasize self-marketing and individualization, rather than collective or cooperative perspectives that can reduce social alienation. Content ranking algorithms are similarly optimized towards a profit-seeking objective to increase the time that users spend on the platform. This ends up reinforcing echo chambers and prioritizes sensational content by creating filter bubbles, thereby reducing diversity and pluralism in the information recommended to the users. Collectivisation in fact becomes harder too with the loss of anonymity and secrecy on such platforms, and imposes threats to the safety of the collectivizing members. When something goes wrong, as with data leaks or unrestricted propagation of misinformation on Facebook, they are dismissed as isolated cases of unintended outcomes rather than systemic problems arising from more fundamental issues of values that were marginalized such as the ones to uphold pluralism, truth, democracy, and equality. Zuboff outlines in detail through the example of the rise in surveillance capitalism of how these harms are hidden from plain sight to avoid regulation. An alignment between companies and governments can lead to the creation of blunt and ineffective regulatory instruments.

Following the neoliberal playbook, technology is also projected as a change-agent to improve the world, in collaboration with the state. An example of such a technology directly aimed at entrenching neoliberalism is the Aadhaar unique identity system in India. It was first positioned as a platform to reduce leakages in the delivery of welfare services (Venkatanarayanan, 2017). Proponents from the private sector marketed the technology by arguing that significant leakages in welfare delivery arose from people using false or duplicate identities to corner welfare benefits, or when deserving people were excluded because they were unable to provide satisfactory identity documents for enrolment. These problems would be reduced if a biometric-based identity was made universally available to the citizens, and then used to authenticate access to the services. However, this argument was incorrect because more leakages happened through other operational processes that Aadhaar could not control (R. Khera, 2017). Further, the introduction of Aadhaar disrupted the delivery channels even of legitimate welfare benefits to the extent that many deserving people were excluded and faced significant hardship. Relentless advocacy by civil society to draw attention to these issues generally faced a deaf ear and Aadhaar has gone on to being declared mandatory for welfare delivery. Despite the integrated digital backbone that Aadhaar has enabled for welfare delivery, grievance redressal mechanisms have remained ineffective and little attention has been placed on leveraging this digital backbone to place accountability and diagnose failures so that citizens can smoothly get access to their entitlements (Gupta et al., 2021). The cumulative effect of these disruptions and redirection of accountability has been termed as a *degenerative outcome* of how digital platforms like Aadhaar are in fact eroding the value of welfare mechanisms altogether (Masiero & Arvidsson, 2021).

This was just the beginning though. The next step was to use the Aadhaar infrastructure to deliver welfare payments directly to the bank accounts of people. A massive drive was conducted in 2014 to open bank accounts for all households, ironically, in many cases, without their knowledge. However, even until now during the pandemic in 2020 the deficiencies of this infrastructure have not been sorted out and led to significant exclusion at multiple steps for people to benefit from emergency cash transfers that were announced during the pandemic (Gupta et al., 2021; Mohan, 2018; Seth et al., 2020). The private sector, of course, benefitted all along from building the supporting information technology ecosystem around Aadhaar and positive externalities for business expansion resulting from an improved financial inclusion in the country.

With an increase in financial inclusion, the Aadhaar evangelists then built a Know Your Customer (KYC) layer on top of Aadhaar, so that any private or public service provider could easily validate personal details of people, such as their address or demographic information (Shashidhar, 2018). This positive externality that was created through public funds and positioned as a public good paved the way for rapid sales of financial products to the people. The hugely disruptive demonetization event of November 2016, in the name of curtailing blank money for formalization of the economy, also provided strong headwinds for the proliferation of digital financial products when cash became scarce even for common transactions (Harriss-White, 2017). Popularly known as fintech (short for financial technology) in the venture capitalist community, the eventual trajectory of these initiatives was unmistakably that of finding new markets for debt fuelled consumption followed by securitization of these loans to expand the market for money capital. With an increased availability of global money capital searching for returns, the newly online poor and middle classes in India present a vast market for formal capital to acquire new customers (Gabor & Brooks, 2016; Mahajan & Navin, 2013). It vilifies usurious informal moneylenders, but in fact only replaces them with faceless and bureaucratic corporations. Coming in many forms, like microfinance, instant loans, crowd lending, etc., this exposes people to new vulnerabilities and predatory lending, often leading to grave distress and suicides, even as any net social development gains remain dubious from such financial lending practices (Radhakrishnan, 2015). Tools of behavioural control built on digital footprints of people, perfected by surveillance capitalism, becomes an added tool in the arsenal of the fintech industry. Clearly, technology developed within the prevailing paradigm of capitalism tends to reproduce dispossession.

Due to privacy concerns and also misuse by several companies of auto-signups for people to services they did not need, the use of Aadhaar based KYC by private corporations is currently restricted. The COVID-19 pandemic has however presented a new opportunity, where Aadhaar has again become the backbone to track vaccine administration and build the digital equivalent of vaccination yellow cards (Gelb & Mukherjee, 2021). This strongly entrenches Aadhaar in the health ecosystem as well, which is contemplating the use of national digital health IDs for easier portability of medical health records, even while larger problems of the availability of doctors and skilled staff goes unaddressed (Neelakantan et al., 2018; Phansalkar, 2020). With layers such as cash transfer, KYC, health records, vaccination records, etc., built on top of Aadhaar, and which are then used by the private sector to create profit-making services, concerns have also been raised about the governance of competition on the platform so that insiders or first movers do not have an unfair advantage (Dharmakumar, 2017). However, the private sector has continued to use similar strategies of first supporting the state to build public good platforms, and then using these platforms for its own advantage. This became obvious with the contact-tracing Aarogya Setu mobile application created during the COVID-19 pandemic. Several telemedicine and teleconsultancy start-ups came together to build Aarogya Setu, and once its use was mandated in several contexts by the government, then the same start-ups tried to ride on the platform to sell their tele-consultancy services to the users (Singh & Ramanathan, 2020).

Digital Public Good Platforms

The Aadhaar system, and layers built upon it for KYC, cash transfer, etc., are projected as digital public goods that have accelerated the development of new innovations at a mass scale, especially for underserved populations. This has been generalized into the concept of *societal platforms*, as the foundational technology layers on which an ecosystem of innovators can build rapidly scalable population-wide applications and services (Nilekani, 2017). A similar idea with putting up digital public good infrastructure for health data forms the basis of the National Digital Health Blueprint proposal (GoI, 2020b). Health provider registries, pharmacies, and personal health records identified by a unique health ID for each individual (with access controlled through a consent architecture), can be accessed by approved application providers to build products related to teleconsultancy, diagnostics, insurance, among others. Identical concepts have been proposed for agriculture, skilling and jobs, and smart cities, where farmer registries or talent profiles or data produced by smart city systems, will be made available to spur innovations (GoI, 2020a, 2020c).

Legitimate concerns have been raised about such systems related to the policies and procedures put in place for data privacy, and the need for clear purpose definitions for data collection for each specific data element (S. Banerjee & Sagar, 2021). While these technical concerns may be solvable if due attention is placed on them, a different critical concern with such digital public good platforms is that only a handful of *insider* players may be able to take advantage of them instead of the platforms serving as a fair and competitive space for any innovator (Neelima, 2020). In fact, most of these platforms have been proposed and built by a common core of people, who are then able to take advantage of their advanced familiarity with the platforms to be the first movers in providing services. Further, the positive externalities contributed by the public-funded platforms may not be fully compensated by the private players. Corporate consultants and revolving door appointments from the corporate sector become the conduits through which a legitimacy for such platforms is created within the government, eventually paving the way for usual capitalistic methods to expand markets, formalize them, and create new ones (N. Hayes & Westrup, 2014; Singh, 2021). Even in Europe, studies commissioned by the European Union similarly recommend such infrastructure to improve the ability of private businesses to innovate (Gansen et al., 2018).

Yet another significant concern is that once the government is convinced, they may adopt coercive means to drive the adoption of these platforms even if their design may not be appropriate or the technology may not be ready. This has been heavily documented in how the adoption of the Aadhaar platform was driven by making it mandatory for various purposes despite its shortcomings (J. Drèze, 2021; Masiero & Bailur, 2021). When people face issues, there is no easy way for them to mitigate the problems because grievance redressal channels are inadequately provisioned, and appropriate laws and procedures to place accountability are also either missing or not enforced (Seth, 2020c). Unconcerned with these gaps, the groups who proposed these systems in the first place resort to deflecting criticism by declaring that the problems arise not from any limitations of the technology itself but from the concerned service provider who did not build safeguarding processes around the use of the technology.

The history of many other technologies in the capitalist context can be traced similarly, through misleading marketing and government lobbying to create adoption, while the technologies further entrench capitalist ideology in society, increase social alienation, and continue the dispossession of the poor and weak at the hands of the elite, through complicity of the state. All the while though, technology is positioned as being useful for society, and corporate controlled media further reinforces this belief to create public acceptance for technology-powered growth without questioning its harms or even the rationality of its purported usefulness. My students examined the media coverage of several government-led technology for development policies in India, and found clear evidence of biased coverage by the media of more positive than negative aspects, with the positive aspects further endorsed by prominent business people from the private sector (A. Sen, Priya et al., 2019).

The attraction of technology to serve as a change-agent goes beyond capitalism alone. Socialist economies have also been enamoured by technologies of industrial production, and as discussed by Graeber, these economies have tended to deploy technologies more for direct use-value emerging from the goals of equity and fairness, rather than profit-seeking directions that emerge through capitalist markets (Graeber, 2015). However, even though these non-capitalist social structures were more oriented towards emancipatory use of technology, the technologies were not always successful. This requires deeper reflection on what forms of technology or regulation are more suitable than others so that technology empowers rather than disempowers people. E. F. Schumacher in *Small is Beautiful: A Study of Economics as if People Mattered* demonstrates many examples of simple and locally produced technologies that fulfil actual use-value for the communities and benefits them, but does not require returns-seeking complex debt or equity structures of capital to finance them (Schumacher, 1973). James Scott in *Seeing Like a State: How Certain Schemes to Improve the Human Condition Have Failed*, gives many examples of both capitalist and socialist inspired plans of using technology in agriculture, urbanization, governance, forestation, etc., that ultimately failed because they were based on overly simplistic models of nature and of technology optimism (Scott, 1998). They did not take into account the complex processes that unfold on the ground through which ordinary people appropriate technology and use their experience to navigate new and emergent situations. Arturo Escobar echoes these views in *Encountering Development: The Making and Unmaking of the Third World*. He attributes the failure of development methods imposed by the Global North to the hegemony of rationalist and modernist approaches that define problems and find solutions without paying adequate attention to understanding the culture, traditions, complexity of social relationships in the Global South, and their entanglement with nature (Escobar, 1995). The myth of technology optimism has however prevailed, even though technology may not always succeed in meeting the stated goals, and may be more oriented to serve to reproduce the social systems within which it exists. Countering this myth becomes essential if technology is to be gainfully deployed for solving the problems facing the world.

Appropriate Technology

E. F. Schumacher distinguished appropriate technology from technologies of mass production (Schumacher, 1973). These technologies have a primary function to enhance the economic position of those who have not been given adequate opportunity to participate in the global development process. They need to be created in workplaces where people are living, so that people are not required to migrate to central hubs of production. The workplaces should be low-cost so that they can be created in large numbers. The production methods should be relatively simple so that demands for high-skills are minimized, and not only in the production process but also for the organization of production, supply of raw material, financing, and marketing. The production should therefore focus on using local materials. Another term used for appropriate technology is *intermediate technology*: which is more productive than the labour intensive and inefficient traditional technologies, but less costly and more manageable than the large scale, labour saving but capital intensive technologies of industrialized societies. Information systems that enable communication between the workplaces is suggested as an intellectual infrastructure to support a move towards intermediate technologies by creating networks of action groups and facilitation centres that enable sharing of know-how and up-skilling.

Schumacher's underlying motivation for appropriate technology came from the sheer disregard of the environment exhibited by the economics and technologies of mass production. These systems did not take *natural capital* into account, that is, capital which is provided by nature. Depletion of natural capital, further accentuated through vices such as the greed of the rich, leads to them prospering at the cost of the poor, by, for example, relocating polluting industries away from themselves to the living spaces of the poor. Schumacher therefore advocated Buddhist economics which aims to conserve and grow natural capital. This requires local production processes for the proper use of land, education of the people for the knowledge and wisdom to operate these processes, and notions of justice aimed at not just lifting the bottom but also lowering the top to create wider cooperation and solidarity.

A similar argument is made by Ivan Illich in his book *Tools for Conviviality* (Illich, 1973). Illich begins with showing that tools may first be made to meet some needs, but soon they create their own problems and require bureaucracies or other means for social control. This reduces the freedom of people and opens up spaces for political control to be imposed over them. In contrast, tools that people can understand and control themselves lead to creativity and foster enriching relationships between people. These are therefore called tools for conviviality, and in contrast with tools of industrial production, they lead to a one-ness among humans and with nature, they slow down the pace of technology development so that it can be sufficiently understood and used responsibly, and they put limits on environmental degradation. Like Schumacher, Illich drew attention to the harms of industrialization and the methods through which its hegemony is established as the only way to make progress in meeting social needs.

2.5 Answers

I use the term technologists to refer to people who build and manage information technologies – the engineers, designers, researchers, and managers – involved with the information technologies we see around us. I identify myself as a technologist too. The discussion so far sets the context for the question I attempt to investigate in this book, of what should technologists do, given the current state of the world where technology seems to be most used in the service of upholding the structures of capital, and technologists are clearly complicit in contributing towards this? (Braverman, 1974). It is indeed computing technologies that today enable the close control and monitoring of workers in most industries, book-keeping of support structures required by capitalism for value accounting and financial management, behavioural predictions for marketing and advertising, among others, and all of these are built and maintained by white-collar technologists. I am confident, though, that most technologists are well-meaning people. We would not like our work to hurt innocent people, we would like it to help others, fulfil a genuine use-value for society, and address the social problems we see around us. Not being able to do good for society is alienating since ultimately we too derive our humanism from being able to see the output of our labour contribute towards positive social relationships that can tie society together into pursuing meaningful goals for one another. In other words, our natural state of being is to do good for society. What then is this social good that can also prevent an alienation among technologists? Can we define it more precisely?

Doing social good should certainly include addressing some of the problems we started out with discussing, including poverty, health, education, global warming, pollution, etc. I have however also discussed how the dominant systems of the state and markets are actually geared towards exploitation and divisiveness, and therefore any solutions for social good built within these systems risk further exacerbating inequality and dispossession of the poor. Technologists who may seem to be in a strong position to do social good, therefore need to be wary since even technologies for social good if left to themselves can become an instrument of oppression and exploitation in the hands of capital.

Controlling capital is however not easy. Capitalism is devoid of any ethical guardrails of its own to govern its power – it can co-opt states into complicity, manipulate markets, use technologies and technologists, and influence consumers. Hence there needs to be some specific form of social good which is non-exploitative and leads to equality, keeps capital in check, and provides guardrails so that technologists do not contribute to building technologies of surveillance and control used by hegemonic powers to entrench themselves further. This is likely to require new directions in which choices are made for reasons of morality and humanism, and not economic rationality; a direction that ensures non-exploitative social relationships emerge from the work while exploitative ones are removed; and that such non-alienating principles become the basis on which the current institutions of governance and the economy are either redone or new systems emerge.

I therefore start with discussing in Chapter 3 when a particular technology project can be said to be doing social good. Is it related to the goals that the project pursues, such as to address the kind of problems we started discussing, of poverty, the environment, etc.? What guardrails should the means towards these ends follow, so that the projects do not disempower people whom they were supposed to support? Should it include wider goals to increase humanism by countering or replacing entrenched institutions of governance and the economy which are disempowering for people? What goals should be included, and who decides what are worthwhile goals to pursue? Is it feasible for all technologists to align their work with these goals and determine whether their labour is leading to social good or not? Chapter 3 provides a framework to define social good and answer these questions. Chapters 4–6 then explain several ways in which harms arise from technology, and ask how technology projects can be designed, managed, and audited to ensure that they are leading to social good. Answering these questions can help technologists choose with which projects they should align themselves, and the ethos they should embrace in their work to ensure that responsible outcomes arise from their labour.

As will become clear in these chapters, doing social good involves going beyond just designing a particular technology artefact, and involves also looking at the wider processes of management, assigning accountability, and understanding outcomes when the artefact is put to use. I refer to all these aspects collectively as being a part of the technology *project* or technology *programme* in which the technology artefact is used. Projects are therefore specific forms in which a given set of technology artefacts may be put to use (Salter, 2007). The same artefacts may be put to different uses in different projects, sometimes leading to social good and at other times to undesirable outcomes. In this book, I therefore do not get into debates about technology determinism or the social shaping of technology which often attempt to generalize arguments about specific technology artefacts (MacKenzie & Wajcman, 1999; M. Smith & L. Marx, 1996). I do not look at artefacts in isolation from how they are deployed and the outcomes that they affect, rather

I consider the technology project as the unit of analysis. My wider message, as will become clear, is that it is important even for technologists who may just be involved in designing a technology artefact, to also be mindful about the projects in which the artefact designed by them is envisioned to be deployed. Only such a consideration can give clarity on whether the technology artefact will lead to social good or not. This can be done for all technology artefacts irrespective of where they are deployed in a value chain in some technology project. If an artefact finds use in multiple projects then its impact can be evaluated in aggregate across all these projects. Henceforth, I use the terms *technology* and *project* interchangeably. Both uses should be interpreted as the same, as examining a technology-based project.

Since capital may not itself choose to pursue genuine social good unless it is mandated by the state or through consumer preferences, I also discuss how, in the current context, technologists can become change agents themselves and can redefine projects in which they are involved, so that the projects can be transformed into genuine social good projects. What mechanisms within their own organizations can technologists adopt to ensure that social good arises from their work? Chapter 7 examines the role of technologists and what constrains them from being able to ensure that their work leads to social good. Chapter 8 proposes strategies such as collective action, use of intellectual property rights, and other methods, through which technologists can gain power within their organizations to govern the use of technologies towards social good.

Projects for social good cannot, however, succeed without the participation of society to endorse these projects. Their use and adoption will depend upon society agreeing with them. As consumers, it is the members of society with paying capacity who would need to define what values they respect, what goals should be considered as constituting social good, and what means of pursuing these goals should be acceptable. Citizens - consumers and non-consumers alike - similarly need to influence their states to uphold the same values in the institutions of the state, and through them to impose social control over technology so that it adheres to the accepted value system. This societal-scale alignment can however only happen if the members of society are not apathetic towards others. How can more empathy be created in society? What projects should technologists support so that society can use this empathy to hold the state more accountable to prevent disempowerment of the weak, to keep a check on capital, and to be more responsible in its own preferences rather than let them be shaped by capital? How can technologists doing social good make society more aware about the risks involved with various technologies and likewise, remain better informed themselves about the impact of technologies designed and managed by them? Chapter 9 examines the systems of media and communication that have a strong influence in shaping society and its values, and I outline some aspects that these systems should follow so that they can contribute to building a more empathetic society and thereby build the ability to impose social control on technology itself.

Rather than provide specific answers though, my main goal is to provide a framework to technologists to be more aware about the implications of their work, and to form a vested interest in governing the use of technologies produced through their labour. I feel that this is not only essential from a rational point of view to ensure that technology does not harm others, but it is also imperative from a moral point of view for technologists to take responsibility of putting their skills and talents towards use for social good, which emerges from an inherent humanism of technologists to connect with others and support them.

The concept of technology for social good is clearly not realizeable in any easy way, but is important to fight for, and this needs to be done on multiple fronts: prevent co-optation of the concept, define it more precisely and legibly, and use it as a means to have technologists, people, and citizens alike to critically examine concepts like technology and social good, and to take appropriate follow-up action. This book is based on the premise that technologists are indeed in a strong position to do social good. If done well, I feel that technology can become an unambiguous means for social good, and technologists can play a strong role in course correcting the trajectory of the world.

Chapter 3

Understanding Social Good

As we head deeper into an information age with increasing connectivity and observability, society seems to be becoming more responsible. We see signs of this with a rise in concepts like stakeholder capitalism, social entrepreneurship, doing social good, and global movements in support of building a sustainable society. However, as discussed in Chapter 2, current social systems pose a risk that in the search of new markets neoliberalism may invent false social problems to solve, or the solutions may continue to be exploitative and dispossess the poor, or the media may shape public opinion to conceal harms arising from these projects. Keeping this in mind, in this Chapter I look at five aspects of doing social good that seem necessary for technologists to understand, to decide how to align their efforts so that they fulfil genuine use-value for society. First, I warn about misuse of the term social good and how we need to guard against the concept getting co-opted by capital. Second, I suggest that reasoning about whether a project is for the social good or not, should be done using the tools of ethics. These tools specifically help to reason about what should be done and why, and provide a distinction between ends and means, which I show is essential in defining social good. Third, I argue that underlying ethical values and ideologies drive all technology and scientific work, and technologists cannot argue for their work to be value neutral. They consciously need to adopt an ethics to justify whether their work is leading to social good or not. Fourth, I point out that doing social good goes beyond just direct attempts to tackle social problems that we have discussed earlier, and requires a total transformation of the systems of governance and the economy itself. Projects with such goals are what I call *meta-social good projects*. Finally, I point out that society needs to democratically participate in deciding the ends and means of what should or should not constitute social good, and that technologists have an opportunity to build such systems for participatory democracy.

3.1 Ambiguity

The concept of social good has over the last decade or so emerged prominently in public conversation, due to an urgency to address problems of inequality,

Technology and (Dis)Empowerment: A Call to Technologists, 33–54

doi:10.1108/978-1-80382-393-520221003

exploitation, global warming, universal access to health and education, and other development concerns. It is further believed that these problems can be solved or existing solutions can be made more efficient through the involvement of technology (Toyama, 2015). Many organizations claim to be using technology for social good to characterize their goals and distinguish their work from typical commercial uses of technology (Unwin, 2017). There is however so far no clear definition of the concept of social good, except for a purposive flavour that it brings about some manner of betterment of society. Being able to clearly define social good can help remove ambiguities and build methods to verify claims of whether a project or initiative is truly doing social good or not. Here I do not mean verification in the sense of impact evaluation of the outcomes arising from various projects, but in the sense of legitimizing claims that organizations make to characterize their work. Verifying claims of the use of technology for social good becomes important because, as I have discussed in the previous chapter, technology tends to reflect the values of the social systems in which it is created and typically gets used to strengthen the hegemony of these systems even further, even if it is popularly positioned as being beneficial to society. It is therefore essential to lay down the conditions under which technology for social good projects are more likely to actually do good, rather than result in further exploitation and dispossession of the weak.

Removing ambiguity in what is social good is also important because the concept seems to be gaining statutory recognition such as with social stock exchanges meant to fund social good projects (Aithala, 2020). In India, any enterprise that "declares an intent to create a social impact and commits to measuring and reporting that impact" can be considered as a social enterprise that can be listed on the social stock exchange. Ambiguity about what is social good can mean that organizations not truly delivering social good may also be able to gain access to such exchanges at the cost of truly genuine initiatives. Simply restricting access based on a for-profit or not-for-profit status of the organizations, or whether or not their work is aimed towards improving standardized indicators such as the Social Development Goals (SDG), would be very crude (SDGs, 2015). Complex arrangements of webs of companies and non-profit foundations can make it hard to draw any clear lines simply based on the tax-status of an organization, and similarly the SDGs themselves are just a few of many quantified sets of categories to which not all organizations may be able to relate their work (Kleine, 2010). A more nuanced method is therefore needed to define social good, and technologies for social good.

Defining the concept clearly may also present an opportunity to draw a distinction between the goals and methods of different organizations. This in turn can guide consumers to choose between competing technologies that may appear to be functionally similar but have different underlying social good goals (Irwin, 2015; K. White et al. 2019). As consumers become more aware of the underlying ethics of work practices as well as outcomes arising from technologies that they use, their choices can shape priorities of corporations to be more ethical, or even create a market incentive for doing social good. Employees within corporations can similarly evaluate whether their labour is leading to social good or not, and internally collectivize to push their companies to do social good (Seth, 2021a).

Unfortunately however due to an ambiguity about what exactly is social good, corporations seem to be co-opting the concept. First, there is a trend towards overloading of the terminology itself. The management literature has lately seen the rise of concepts like purpose-driven organizations and that corporations should ensure social justice, which appear similar to doing social good (Mourkogiannis, 2014; New School Series, 2020). A deeper examination reveals that these concepts are however not advocating for any shift in the goals of corporations, such as towards meeting some social development goals. They are rather espousing internal operational values such as non-discrimination in employee recruitment and growth, and professional conduct for values like excellence and innovation. These are important but do not advance the concept of corporate responsibility or corporate governance beyond what is even legally deemed essential in most countries now. Merely following basic rules like non-discrimination or aspiring for excellence in work, cannot qualify as social good.

Second, the concept of doing social good seems to increasingly be used as a deflective or conscience laundering technique while corporations otherwise go about their business as usual. This has been noticed prominently with themes like AI4Good being actively promoted by companies, such as Google and Facebook, and the notion of *supercorps* – vanguard companies that are able to repurpose their organizational capacity to solve important social problems (Kanter, 2010). Especially during the pandemic, governments too seem to have started relying on the capacity of these corporations, such as for contact tracing and mass communication (Iazzolino et al., 2020). Propaganda of doing social good could cover-up the harms, even if unintended, that the work of such corporations otherwise brings about.

Third, a distinction that broadly exists between social enterprises and commercial enterprises can vanish if both claim to be doing social good. An analysis of the mission statements of several technology companies showed that they all claim to be empowering people in some way: Microsoft to *empower every person … to achieve more*, Facebook to *give people the power to build community …*, Twitter to *give people the power to create and share ideas …*, Tumblr to *empower creators …*, etc (H. Schneider et al., 2018). A method is needed to decide which of these should be called social good, otherwise the concept could lose its value and distinctiveness.

Fourth, while social good has typically included social development to solve problems such as inequality and ensuring universal health access, the term development itself is highly contested on whether it implies economic development or social development or equates the two, and can add to concerns of co-optation of the concept of social good (A. Sen, 2000). A non-technological example (although with similar characteristics as naive technology optimism) is microfinance. Originally projected as a *packaged intervention* that eliminates usurious informal money-lenders and thereby provides easy access to credit for the poor, studies indicate that no net social development gains actually come about from microfinance (A. Banerjee et al., 2015; Toyama, 2015). Simply due to a formalization of the credit market of the poor, more capital is able to enter the system, making consumerism easier and thereby at best boosting economic growth.

Despite this evidence that microfinance potentially brings about economic development but negligible social development, it still continues to be considered as a prominent example of social entrepreneurship. There is little in fact that distinguishes contemporary microfinance from commercial lending, apart from the customer base being that of poorer people and the operational mechanisms being different to service this demographic. This extends to many ICTs for development interventions, where, for example, mobile applications with no clear path to social development regularly receive impact funding from investors in the name of doing social good just because these applications are meant to service a new user base that was not using ICTs so far (Seth, 2020d). A confusion about social good therefore prevails even in the social entrepreneurship space, of whether these projects are merely expanding the market for capital or are they indeed empowering the poor (Chell et al., 2016). For example, many social entrepreneurial projects in public health or education are aimed towards the privatization of public services and reduce the state's responsibility for social development, by providing welfare services for a fee or through sponsorship by the state. Such projects tend to weaken the citizen–state interface which then impacts democracy and weakens community institutions to demand accountability directly from the state.

Fifth, doing social good is not simple. As I outline through many examples in the forthcoming sections and chapters, competing priorities of technology projects and restricted choices available to people in diverse social contexts can result in widely different outcomes that may or may not solve the given problem or empower the people that the projects intended to support. Without a strong concerted effort to design and manage technology projects towards specific outcomes, prevailing paradigms of technology design and management tend to reproduce inequality with their narrower focus on self-interest and economic gain that favours those already in power. A clear definition for social good is therefore needed to unambiguously categorize actions and outcomes as whether they are consistent with the definition or not.

The motivation to resolve some of these concerns is similar to that raised by Ekbia and Nardi who recognize the inclination of corporations to be driven by financial gain instead of values (Ekbia & Nardi, 2015). They suggest making explicit who benefits and who is left behind from technology and its outcomes. My proposed method can be considered as a way to move forward in this direction.

Microfinance in Practice

Poor families, especially those that draw part of their income from agriculture, do not have a regular cash flow (Collins et al., 2009). Health-related emergencies, or failed crops, or other exigencies, can therefore throw families into poverty, especially when safety nets are not available or operate well (Krishna, 2017). Access to formal credit can certainly help in such situations, as it can help poor households to start new enterprises or diversify their income sources at lower costs than borrowing from informal sources (Morduch & Haley, 2002). Yet several studies have shown that microfinance does not bring any net benefits, or only small gains at best depending upon the local context (A. Banerjee et al., 2015; Gopalaswamy et al., 2016). While microfinance is extolled as an anti-poverty tool that supports entrepreneurship, especially among women by contributing towards gender equality, in reality it may not even reach the poor or be used to set up enterprises or enable women to take independent decisions (Radhakrishnan, 2015).

From Gram Vaani's own experience of working on the ground in affiliation with several Microfinance Institutions (MFIs) in India, we can appreciate the opportunities offered by the model but also its blind spots. We saw on the ground that it is indeed hard to reach the poor. Loan officers from block level offices of MFIs leave each morning on their motorbikes to make fortnightly or monthly collections from the hundreds of joint liability groups of women borrowers in the area. With daily targets against which their performance is appraised, the loan officers have to rush from one group to the next, with no time to offer any reasonable advice to anybody facing repayment issues or requiring help with their enterprises. The same households are shortlisted by multiple MFIs operating in the area as being low-risk borrowers, and multiple lending is common despite regulatory norms against it. Many other poorer households who need loans and apply for them, are regularly denied access. Several borrowers do not use the funds for any income enhancing activities, but rather to smooth their cash flow. Many borrowers experience stress to meet the regular repayment terms, and mandatory capacity building workshops to help people handle the microfinance process are conducted summarily only to meet compliance requirements. Any actual capacity building is done through grant-funded projects at small scales, since MFIs do not have enough profit margins to deploy resources outside of running their mechanistic operations of lending and collection.

As part of one of our projects, there were clear differences we noticed between borrowers from a commercial MFI, and those from a hybrid organization involved in social development work along with microfinance. The greater impact of the hybrid MFI was clearly visible – it had first mentored women to take up entrepreneurship, engaged in wider projects in the community on social norms related to gender equality, and the enterprises then created by women using MFI borrowing were more successful than the ones spun off using loans from commercial MFIs.

In short, as I reiterate throughout this book, it is not easy to do social good. It is a complex process, which, if over-simplified just to make the concept attractive and marketable as an anti-poverty silver bullet, can rather hurt the people it was meant to support. Even worse, it may result in a wealth transfer from the poor to the rich. The complexity of doing social good should rather be acknowledged, and instead of seeing social entrepreneurship as a means to address market gaps it should be seen as a method to serve people who cannot be served through the market in the first place. A substitute for welfare cannot be found in the market, rather uplifting the poor or bringing equality requires redistribution mechanisms that are executed with efficiency and ingenuity. That is true social entrepreneurship.

3.2 Resolving Ambiguities Through Ethics-based Methods

Assuming that a clear definition of technology for social good will help create a distinction from other technologies, and improve the efficacy of mechanisms to support the growth of such technologies, the question is: When can a technology be legitimately claimed that it is meant for social good?

I first show that ethics presents a rich diversity of values or systems of thinking that can be used to construct definitions of what is social good. Ethical frameworks of consequentialism, deontological ethics, and virtue ethics are specifically meant to answer questions about what is right and why. I do not suggest adopting any one definition, rather an ethics-based framing is particularly well suited to define social good because it can accommodate a large variety of definitions and provide principles to reason about when something is or is not aligned with the given definition so as to be called ethical. A definition of social good should at the very least be ethical according to some ethical framework.

I argue further that doing social good is essentially purposive and therefore the ethics-based formulation for social good should specify consequentialist ends and not just means. Simply defining ethical guidelines of how a technology system should function is not enough for it to bring about social good, rather an end-point should be clearly specified using an ethics-based formulation. I use this to indicate that only adhering to ethical codes of conduct written by various professional bodies, companies, and even governments, and including statements on Artificial Intelligence (AI) ethics, cannot be considered as sufficient to claim social good because these plethora of statements do not clearly specify any end-goal (ACM, 2017, 2020; IEEE, 2014). Such guidelines are not sufficient to steer projects so that they can unambiguously lead to the benefit of society. In other words, the ethical value of non-malfeasance to *do no harm* is not adequate in itself; it requires a value of beneficence to lead to positive outcomes.

Additionally, specific methods would be needed to identify the values that can guide what ends and means are worthwhile to pursue. I suggest that choosing these values should be done through participatory and democratic procedures in society. These choices should be adopted both in the governance of markets in which society members participate as consumers, and in the values accepted by the state in which society members participate as citizens. Although not all citizens may be consumers, but democracy is essential for them to understand one another, and then enable consumers to shape markets through their preferences and citizens to shape policies and regulations imposed by the state. Finally, project designers and managers should take responsibility to ensure that these values deemed important by consumers and citizens for their projects are indeed realized in practice and reported publicly.

3.2.1 Expressing Social Good in an Ethics-based Terminology

A systematic analysis of the practice of *social good*, that is, how the term is currently used in multidisciplinary academic literature, the web, and interviews of experts, reveals three domains that social good projects typically address: diversity and inclusion, environmental justice and sustainability, and peace and collaboration (Barak, 2020). The first domain draws attention towards preventing exclusion on the lines of gender, race, caste, and class in healthcare services, educational attainment, clean water, ensuring food security, and employment, among others. The second domain advocates for fair access to resources, intergenerational justice, and to avoid disproportional negative impact on vulnerable populations. The third domain emphasizes to prevent conflict that arises on religious or national grounds, to espouse a global perspective that embraces pluralism rather than a narrow community-centric localized view that negates other perspectives. All these three domains of social good are purposive in nature, targeted at particular goals, and can clearly be related to well-known ethical concepts such as by Jeremy Bentham, John Stuart Mills, John Rawls, and Immanuel Kant (see box on *ethical theories*). Preventing exclusion is related to ensuring *equality* in access to essential services and basic needs, intergenerational justice and protection of vulnerable groups is related to principles of *fairness* and *consequentialism*, and reducing conflict arising from cultural relativist arguments is related to respecting the *rationality of others* and *universalizing values* of not harming others. It should be possible therefore to express social good in terms of ethical concepts. Doing this can be quite useful, as we outline next.

Ethical Theories

Different perspectives have emerged over the years of looking at the ethics of computer technologies. Machine ethics delegates responsibility to the machine by making it ethics-aware so that it can reason and decide actions based on computationalized models of ethical frameworks (Allen et al., 2006; Moor, 2006). Another view considers that the responsibility should lie with humans who build and design technologies, including for the appropriateness of decisions made by algorithms that operate inside the machines (Rogerson, 2010). These designers and developers of technologies need to reason about what technologies to build (Weizenbaum, 1976), foresee any policy vacuums that might arise with new technologies (Moor, 1985), and put checks and balances into place so that existing problems are not aggravated with the introduction of computer technologies (Bynum, 2000). In this book, I largely focus on this second perspective, to understand what goals technology should be designed and scaled to address, and how to verify whether progress is being made towards these goals or not.

The field of ethics provides useful frameworks to reason what should humans do and why. Three such broad frameworks that are commonly discussed are consequentialism, deontological ethics, and virtue ethics (Quinn, 2014; M. Sandel, 2009).

Consequentialism advocates for consideration of the consequences of a decision in terms of the welfare of all stakeholders impacted by the decision. Obvious problems with operationalizing the framework arise from being able to define what is welfare, who are the stakeholders to consider, how to evaluate the welfare of one stakeholder against that of another, and how to factor uncertainty in the consequences of an action or rule. The contribution of consequentialism however lies in acknowledging that all stakeholders should be considered, and the consequences of an action or rule should be examined, which provides an important lens when thinking about what technologies to build and why. Utilitarianism of Jeremy Bentham, and improvements made upon it by John Stuart Mill, such as by distinguishing between higher and lower forms of welfare, have had wide influence in fields like economics.

Deontological ethics draws attention to universal rules that people should abide as a duty or as part of a social contract. Immanuel Kant's categorical imperative identifies as good only those acts emerging from moral rules of a person if the same rules were followed by everybody else as well. Its significance lies in considering everybody as equal and respecting their rationality and autonomy. John Rawls built similar ideas into a theory of justice as following two principles. With an emphasis on equality, the first principle states that a person may claim certain rights only if everybody else can claim them as well. With an emphasis on fairness, the second principle states that any social and economic inequality can be justified only if it has arisen after having provided equal opportunity to everybody else, and that these inequalities should produce the greatest benefits for the least-advantaged members of society. Deontological ethics therefore provides valuable tools to think in terms of universalism and fairness, such as Rawls' *veil of ignorance* which advises that a person making some decision should place themselves in the shoes of unspecified people who would be affected by the decision, thus neutralizing any impact of their own power or position in the decision-making process. It also emphasizes on a social contract that requires the welfare of the poor, and in guaranteeing certain basic needs and opportunities as essential rights.

Both consequentialism and deontological ethics are based on objective criteria to govern decision-making and can sometimes fail to produce what might be obviously reasonable results. Several case studies of dilemmas point to the need to account for morality and emotion in decision-making as well. Rather than coming up with objective rules, virtue ethics proposes decision-making to be measured against an ideal of what a virtuous person would do in the same situation. It acknowledges that with more wisdom and experience people can become more virtuous over time, and highlights the importance of virtues, such as honesty, courage, compassion, generosity, integrity, fairness, self-control, among others. The history of virtue ethics goes all the way back to Aristotle, who distinguished between two kinds of virtues: Intellectual virtues associated with reasoning and truth, and moral virtues associated with the character of a person and formed through habit and experience over time. With a focus on these moral virtues, virtue ethics allows for different conceptions of virtues among people which makes it limited in identifying universal norms, but powerful in being able to understand the reasons behind an action based on the specific set of virtues and priorities that were followed in the decision-making process at that time.

These three frameworks provide valuable lenses through which technologies and technology projects can be examined, such as consequentialism's emphasis on consequences, considering all stakeholders, and long-term and unintended effects; deontological ethics' emphasis on autonomy, dignity, rights, duties, justice, fairness, transparency, and universalism; and virtue ethics' emphasis on role models, evolving wisdom, and setting the right example for others (Vallor et al., 2018). They can help construct definitions of *what* is social good and *why* by providing a reasoning framework. I do not suggest adopting any one framework, but that different frameworks can provide different lenses through which a definition for social good can be examined and whether a project is aligned with this definitions or not. I also suggest that participatory democratic methods should be used to decide what social good goals should be followed by society.

I further argue in this chapter that social good is purposive and consequentialist, meant to achieve particular goals. This can be used to select ethical frameworks that explicitly allow end-goals to be modelled, beyond just a set of means to be followed which may or may not help meet the goals.

To explain by way of an example, consider a hypothetical mobile phone-based technology project that provides to farmers the latest prices in different marketyards so that the farmers can make informed decisions of where to sell their produce. Can this project qualify as a case of using technology for social good? An ethics-based analysis can be used to reason this. Having information about market prices can be considered in this case as a basic need for farmers, to avoid information asymmetries, and hence the project would meet the social good criteria of reducing exclusion from access to this basic need. To meet this basic need however, all farmers should be able to afford mobile phones, subscribe to the market prices service, and possess the know-how to use the service. One perspective can be that owning mobile phones and knowing how to operate them and to use the service, should therefore also be included in the set of basic needs. To truly qualify as a social good project, the technology provider should fulfil these needs by also training the farmers and purchasing phones for them, or facilitate access through other means like providing credit for them to purchase phones. Rawls considers having access to basic needs as a form of liberty, and that everybody should have an equal opportunity of access to venues where they can use these liberties (M. Sandel, 2009). A Rawlsian analysis would indicate that indeed the technology provider should go beyond just making market prices available, so that all participant farmers have the necessary resources to get the pricing information and an equal opportunity to use it.

Inequality may thereon still arise among the farmers based on their relative capability to use the information. Some farmers for instance may be smarter than others and will manage to use the same information to additionally even forecast future market prices and factor those in their marketing decisions. This emergent inequality would be permitted in Rawls framework since it would not have arisen from any inequality in the opportunity to use the information. The technology provider may go further and facilitate reselling this price forecasting capability of the smarter farmers to other farmers, by collecting a value-added service fees from the farmers and splitting it with the smarter farmers, potentially based on the accuracy of their price forecasting capabilities. The larger gains that would accrue to both the technology provider and the smarter farmers would also be permitted in Rawls framework through the justification that it actually makes all farmers better off by further strengthening their bargaining power in the market.

However, a different notion of distributive justice that emphasizes equality may yield a different answer to how the technology provider should scope the project, such as suggesting taxes on the unequal benefits gained by different farmers so as to bring more equality among them. A consequentialist view of a fear that the price forecasting value-added service could degenerate into a marketplace for speculators may also lead to a decision that the technology provider should not support such a service.

Different ethics-based foundations may therefore yield different answers to what would qualify as a technology for social good project. The inherent design of the technology, the scope to which the implementing organization may restrict itself, the stakeholders it empowers and those it leaves behind, its compatibility with the wider accepted norms in society, can give different answers based on how social good is defined. I am not suggesting any particular ethical system or a list of values or a particular ranking of values that should be used to define social good, but that when an ethics-based foundation is used then various principled tools and methods developed in the field of ethics can be employed to reason whether a technology is ethical, that is, in line with the ethics-based foundations of how social good is defined. This can be taken further to also assign accountability to the technology designers and managers, in determining the scope of their efforts and responsibilities to ensure that the project leads to social good.

Several ethical frameworks can be used for this purpose. Value sensitive design (VSD) is one such ethics-based framework for technology projects. It lists 18 values, and suggests several tools to resolve conflicts among these values through a discussion among various stakeholders involved in the technology design (Friedman et al., 2013). Others like the Ethical Design Toolkit similarly list 22 values evaluated on the three ethical frameworks (virtue ethics, deontological ethics, and consequentialism) that technologists can discuss and incorporate in their technologies (Ethical Design Toolkit, 2020). The capabilities approach similarly lists different types of freedoms to realize capabilities, and provides many examples of how multiple freedoms are interdependent on one another (A. Sen, 2000). The absence of freedoms to realize some capabilities can prevent individuals from accomplishing what they may desire. Some of these capabilities can even be deemed as basic rights, sometimes guaranteed by law enacted in a democratic manner, to define a social contract that technology providers would be required to comply with (Gasper, 2004; A. Sen, 2000). A failure to comply with the social contract would automatically amount to disqualification from a claim of doing social good. The Rawlsian and Kantian frameworks serve to define universal social contracts based on their respective principles, and can help reason out the compliance of social good projects with these social contracts (M. Sandel, 2009).

3.2.2 The Consequentialist Nature of Doing Social Good

Social good projects are aimed in a purposive manner towards some of the domains as listed earlier, such as to prevent exclusion from access to basic needs, bring about fairness towards vulnerable groups, and reduce conflict due to diversity. When mapped to the space of ethics, the presence of such end-goals implies that the corresponding ethical framing of these end-goals also needs to be of a consequentialist nature. In the area of moral psychology, Rokeach introduced a distinction in human values, between instrumental values and terminal values (Rokeach, 1973). Terminal values refer to desirable and end-states of existence, such as equality, a world at peace, freedom, and welfare of others. Instrumental values refer to preferable modes of behaviour as a means to achieve the terminal values, and include honesty, politeness, responsibility, and sustainability. Terminal values are consequentialist, and an ethics-based expression of social good will therefore need to include such terminal values.

Similar to Rokeach, Sen also distinguishes between constitutive freedoms and instrumental freedoms for development (A. Sen, 2000). Constitutive freedoms are those that need no further justification, that is, they are constitutive of development itself and therefore are end-goals, such as freedom from starvation, from illiteracy, and freedom for political participation. Instrumental freedoms are the means to achieve constitutive freedoms, such as the freedom to participate in economic markets, live a healthy life, and have the freedom to scrutinize and criticize authorities. A definition of social good based on Sen's concept of development would therefore include corresponding constitutive freedoms as an essential element in its ethics-based formulation. In other words, if an ethics-based definition of social good does not include any constitutive freedoms like those specified by Sen, or any terminal values like those specified by Rokeach, then it should not be considered as social good.

Since reasoning about social good requires to create a distinction between ends and means, it then raises the question that which ethical frameworks allow such a distinction, and would therefore be suitable to express social good?

VSD, which identifies specific values that should be incorporated in a technology from the outset, does not natively make any instrumental or terminal distinction between its values (and in fact has been criticized for this reason, but this does not mean that VSD cannot accommodate a ranked set of values based on the desired consequentialist definition of social good, Manders-Huits, 2011). Its list of 18 values already spans many social good domains and can be extended further, although there has also been criticism that no clear methodology for expansion of the set of values has been specified in VSD (Dantec et al., 2009).

The Rawlsian framework is somewhat restrictive in maintaining a distinction between ends and means. It does allow that some social good end-goals can be specified as basic liberties that should be available to everybody, but only ensures fairness guarantees for other opportunities. This has drawn Sen's criticism that ensuring fairness alone does not specify what outcomes or social realizations will finally emerge, or how to reason about choices to govern subsequent decisions that may need to be made when the social contract is in play (A. Sen, 2009). Sen points out that the set of freedoms considered as basic freedoms in any given social contract should be decided through discussion among an open and unbounded group of participants, and even whether to rank liberty over equality or equity (as Rawls does, with his first principle operating in precedence over the second principle) should be open to discussion. The situation is similar to that of a market, where simply having the freedom to participate and transact does not say anything about what the market would be used for or where it will take the world. Rawls framework therefore is not complete in itself to define specific consequentialist ends. Sen also suggests that unique and absolute answers that are provided by Rawlsian or utilitarian frameworks may not be relevant - rather, it would be sufficient to evaluate plausible freedom alternatives against one another to identify the most appropriate one suited to the context at that point in time.

Other than the Rawlsian framework however, it still seems feasible to use VSD and the capabilities approach to define social good using an ethics-based formulation, with terminal values or end-goals being essential components of the definition of social good. In contrast, many codes of conduct and ethical statements by companies which are also modelled on similar ethical frameworks, do not clearly state any terminal values or end-goals. An analysis of AI values statements and manifestos issued by several independent institutions showed that they were closer to conventional business ethics rather than aiming for any specific forms of social justice (Greene et al., 2019). The emphasis was on instrumental values such as *accuracy*, *computationally rigorous methods*, engagement of experts from among *different stakeholders*, *responsibility* by *developers and product managers*, and at best vague end-goals like to *make our societies better*. This absence of any clearly stated consequentialist positions to aspire for, I argue, should be factored as not qualifying for social good.

Other analyses have made similar observations that ethical codes of AI companies need to move from business values to also specifying a mission to society (Washington & Kuo, 2020). An analysis of mission statements of 250 companies from across 10 countries revealed that most statements were about relating to customers, and emphasized only on instrumental values of leadership and following ethical business practices, rather than any terminal values (D. King et al., 2014). Terminal values were stated more clearly in corporate social responsibilities (CSR) activities, or social entrepreneurship projects, to address social needs that especially arise from market failures (Farcane et al., 2019). For example, an analysis of CSR activities as mentioned on the websites of Fortune-500 companies, highlighted a clear focus on community welfare and environmental goals (K. Smith & Alexander, 2013). Social entrepreneurs similarly distinguished themselves from business entrepreneurs with clear goals to fulfil social needs (Zahra et al., 2009).

Once clear end-goals are defined, they can then be mapped to an ethics-based formulation of terminal values if using VSD or of constitutive freedoms if using the capability approach. Instrumental values cannot alone guide towards end-goals and would amount to running a rudderless ship. They can act as guardrails but a potentially infinite set of outcomes could emerge and hence terminal values are needed to know where to steer the ship. An ethics-based formulation without any such terminal or end-goals cannot qualify as social good.

Applying this principle to technology for social good claims can help dismiss cases where instrumental values like privacy or trust or do no harm are stated as ends in themselves. A social good claim would need a terminal specification of *privacy and trust to do what? Being responsible* similarly is not enough, responsibility itself requires consequentialist thinking and needs clarity on *responsibility to ensure what?* These instrumental values can serve to achieve terminal values such as achieving an equality of access to some resource, or ensuring universal security of some basic need, or preventing conflict of some form, but cannot provide a definitive direction all by themselves. *Neutral* technologies that do not have any explicitly stated terminal goals should not be considered as supporting social good. Brian Arthur makes a similar argument. He discusses the technology evolution process in *The Nature of Technology*, of how new technologies build upon existing technology components and lead to waves of innovation as new technology domains displace older ones (Arthur, 2009). He outlines however that this process of putting existing components together to build increasingly complex new components is driven by intentions. The innovations are first conceptualized in the minds of the innovators to solve certain given problems, and then fine-tuned and mastered to lead to predictable results. Similarly, scientific research too is argued to not be neutral, inevitable, or accidental – the bulk of science is done for specific and deliberate ends which require funding and political approval, and are thereby shaped by the prevailing paradigms in society (S. Rose & H. Rose, 1973). While there are many stories of serendipitous discovery and innovation that can be argued as leading to *incidental* social good rather than *purposive* social good, these are more of an exception than the norm. Technology development and scientific research is therefore not driven by instrumental values alone – a closer inspection will reveal underlying terminal values against which then social good will have to be justified. This is also true for technologies which may not directly impact users and are deployed upstream in a production value chain. Such technologies can be evaluated based on the final use-values that are met by the different projects in which they are used.

There are many examples where technologists have realized their inescapability from acknowledging the terminal values of the technologies built by them, which persuaded them to take steps to control the use of their inventions. Norbert Wiener refused to give access to the military to several of his research studies on missile control (Wiener, 1950). The Russell–Einstein Manifesto was drawn up by scientists in 1955 to highlight the dangers of nuclear war, and led to the Pugwash conferences which were later instrumental in nuclear test-ban and non-proliferation treaties (Russell Einstein Manifesto, 1955). The Asilomar conference in 1975 to arrive at guidelines for further research on recombinant DNA is another example of the scientists leading the effort for eventual regulation of the responsible use of their discoveries and innovations (Berg, 2008). This conference is also significant for the public engagement model that it followed, to involve the civil society in understanding the technology and its ramifications, and still evokes positive memories among the participants of how scientists were able to successfully acquire public trust and facilitate appropriate regulations through deliberation among various stakeholders (Gisler & Kurath, 2011).

To summarize, the values of innovators and scientists drive the process of technology innovation towards specific purposes, these values manifest themselves in the design and management of technologies, and therefore technologies can be defined as being motivated by social good or not based on these underlying values. When the purpose of technologies is unstated, especially to hide purposes that society is likely to reject if the purposes are disclosed, or when the technology use may lead to unanticipated outcomes that are undesirable, then it is often assumed that regulations and consumer awareness will realign the values of the technology with what is acceptable to society. This however is a mistaken assumption. Zuboff, for example, describes in detail the insidious rise of surveillance capitalism and how it went undetected and unchallenged until it was strongly entrenched in market-based institutions along with having also found alignments with priorities of the state (Zuboff, 2018). The terminal goals of profit seeking through behaviour prediction and ultimately behaviour control, as well as the absence of instrumental values of preserving user autonomy, remained not just unstated but other values of user convenience were prominently highlighted to distract people and habituate them to technologies which in fact were disempowering for them. The underlying values of technologies, including their terminal values, should therefore be formally defined, or even if the pathways towards the end-goals are not entirely clear at the outset, the values behind these hypotheses can help uncover an alignment between the technology and any social good claims that it makes (Merton, 1936).

A critical examination of the values behind technology projects can thus help determine which ones to include or exclude, how to rank them, check for compatibility, and choose between potentially conflicting values such as competitive versus cooperative or individual versus collective values (Hanel et al., 2018). Putting down the value system for a project will also remove ambiguities by making its ethics-based foundation legible and transparent. This will further allow for a wider participation of stakeholders in discussing the values, including consumers and employees, and support more informed decision-making through discussion and deliberation (A. Sen, 2009).

The importance of distinguishing between instrumental and terminal values has not been widely recognized in the technology design literature, except in the Scandinavian participatory design tradition (J. Gregory, 2003), the neo-humanist paradigm in information systems design (Hirschheim & Klein, 1989), and the ETHICS methodology for information systems (Mumford & Weir, 1979). Participatory design uses democracy as its underlying value, both as a means and an end, and consequently upholds values of democracy like equality in participation, decision-making through consensus, capacity building to ensure that all stakeholders have an equal opportunity to participate meaningfully, and empowerment of the weak. The neo-humanist paradigm and the ETHICS methodology both emphasize on emancipation of the weak as a goal, by removing the causes of power differentials between different stakeholders. Neither of these frameworks however explicitly related themselves with ethics. Another framework, Definitions, Issues, Options, Decisions, and Explanations (DIODE) evaluates the design of a technology system in terms of human rights, and can be considered similar to our approach of evaluating in terms of ethical values, although it does not make any specific distinction between instrumental and terminal values (Rogerson, 2017).

3.3 Choosing Values

If an ethics-based formulation of a technology project does espouse terminal values or constitutive freedoms as end-goals, for which values or freedoms can it be considered as a social good project? Clearly only a subset of terminal and instrumental values should be allowed to qualify for an ethics-based definition of social good. Here, while it may be tempting to simply choose end-goals and means from among the UN Human Rights Declaration or the Social Development Goals or to combine the two (SDGs, 2015; UDHR, 1948; Williams & Blaiklock, 2015), I discussed in the previous chapter that many purportedly socially beneficial initiatives may actually fail in benefitting society, and rather serve to entrench further the current exploitative systems of the state and markets. Examples like microfinance, financial inclusion, smart cities, biometric-based tracking, etc., can be stated in terms of strong end-goals such as to reduce poverty, enable access to opportunities for growth, counter global warming, and provide security. However, many forms of these initiatives actually exclusively expand markets for capital, often disregard the local context in their effort to build rapidly scalable systems, do not ensure inclusivity in their usage, and ultimately hurt the poor and effectively increase inequality through wealth transfer to the owners of capital. Defining social good should therefore offer some methods to disqualify such projects or allow them only with additional safeguards.

Marx's approach of humanism offers some guidance here about avoiding such exploitative outcomes by defining the most natural and desirable state of being for humans. For Marx, people derive their humanism from positive social relationships they create through the processes of production. These relationships are not necessarily economic. Non-economic relations of care and nurture, which cannot be traded in the market, are also production relations (Gorz, 1998). The criteria for humanism are for these relations to fulfil genuine use-values in non-coercive ways, without oppression or exploitation of others. This draws attention to both ends and means. Ends that fulfil genuine use-values for society need to be accompanied by means that are non-coercive, do not require a domination of others or their instrumental use, and do not disempower them.

Marxist humanism provides an appropriate direction to discover what is social good. First, with its focus on both means and ends it provides a suitable framework to accommodate terminal and instrumental values that should guide the actions of people during the production process. Second, Marxist humanism coincides with several other conceptualizations of justice that reject exploitation and domination, and embrace values for social good as actions that help others. For example, it can be related to Amartya Sen's capabilities approach which emphasizes freedoms that prevent oppression and domination of people from realizing their capabilities (A. Sen, 2000). Iris Marion Young similarly states that removal of social processes that lead to the oppression and domination of marginalized social groups is a stronger goal than to just ensure an equitable distribution of material resources

(Young, 2011). Third, Marxist humanism being based on relationships created by material and non-material production processes, is applicable just as much to technologists as to other workers or members of society. Fourth, since Marxist humanism sees society as being formed from a result of humans engaging with one another in production processes, it also provides a framework to understand how the institutions of the state and market are created out of these very processes (C. Fuchs, 2020). This is useful because it provides an overarching framework of critical theory to understand social systems described in the previous chapter in terms of employer–worker relationships, corporate–government relationships, citizen–state relationships, among others, about when these relationships may be leading to exploitative or disempowering effects. Marxist humanism therefore provides an underlying basis to Marxist philosophy, relevant tools such as critical theory to understand social systems, a concrete direction to define what is social good in terms of means and ends, and an ethos for technologists to follow in their work.

Critical theory is especially useful to help uncover exploitative or dominating relationships and structures, the reasons behind them, and how to overcome them (Freire, 1970; Gramsci, 1971). Other frameworks may not allow for such inspection and introspection. The aspiration framework, for example, does not offer a guidance on how to prioritize terminal goals that prevent disempowerment and are therefore more worthy to aspire for than other goals (K. Toyama, 2017). Sen's capabilities approach also does not clearly acknowledge the exploitative nature of capital and un-free markets, and therefore does not help in developing strategies to counter them. The rights-based approach defines some fundamental rights that can be mapped to corresponding terminal goals or freedoms, but an underlying ethics on how to select these rights is absent (A. Sen, 2009). Critical theory, on the other hand, provides a lens through which patterns of exploitation and domination can be uncovered, understood in terms of how they arise and operate, and thereby can be used to define rights or terminal goals to address these patterns of disempowerment. Countering these exploitative patterns is essential to build a humanist society.

Social good can therefore be defined using critical theory to identify those values that increase humanism in society by connecting people, including technologists, with one another through social relationships of production that create genuine use-value. Discovering and embedding these values as new paradigms into existing systems of the state and markets, however, cannot be done without the wider participation of society. This societal participation is needed at two levels. First, society has developed for itself mechanisms such as the democratic state and free markets, and methods exist for society to hold these systems accountable so that they adhere to the values deemed essential by the members of society. In theory then, as suggested by Sen, society can democratically debate and agree upon the key freedoms or values that it espouses for social good (A. Sen, 2009). Accordingly, through appropriate regulations enforced by the state, and preferences by consumers, capitalist markets that otherwise do not have any inbuilt values can be made to adhere to essential values chosen by society. For example, markets may be forced to accommodate instrumental values such as to account for environmental externalities, or to rigorously follow labour standards in production processes. Markets may even be made to adhere to some specific terminal values, as in a corporatist or socialist state. Even if markets cannot be forced to restrict themselves to particular terminal values, a recognition of this limitation can itself form the basis to distinguish social good projects from the rest. Social good projects will then be those that aim for the specific terminal values that were decided through democratic means as those that increase humanism in society. Projects which are aimed at these relevant terminal values can then be allocated funding or state support to mobilize them.

Second, while the above setup may appear complete, I have discussed earlier that many mechanisms through which society enforces accountability on states and free markets are actually corrupted and broken. A category of *meta-social good projects* is therefore needed to fix these broken accountability mechanisms and transform the prevailing systems of the state and markets. This would include projects such as the following:

- Participatory projects that help societies to discover and agree upon essential terminal and instrumental values for society members to follow.
- Projects that facilitate society members in the governance of markets to ensure that these values are followed in the market, so as to eventually alter the values of capitalism itself.
- Projects that facilitate the governance of the state in making it accountable to society so that the selected values are followed in state-run projects as well, and in markets that are regulated by the state.
- Projects that challenge the hegemony of capital, to either force it to accommodate these new values, or to replace it with some other system.

These meta-social good projects can thus clarify the terminal and instrumental values essential to be considered as social good, transform the systems of the world to enable society to pursue these values, and pave the path for more direct social good projects. Meta-social good projects therefore help establish the conditions within which social good projects can operate and work according to agreed upon values.

What core values do these meta-social good projects need to adhere to so that they can in turn democratically define social good that increases humanism in the social relationships of production that bring people together in society, and strengthen the institutions required to uphold these values? I first propose a framework in Chapters 4 and 5 to minimize harms or unintended undesirable outcomes that may emerge from technologies. The framework can be used to make some quick rejections of projects claiming to do social good, if internally the values of these projects are not aligned with one another, or the projects have no clear terminal values defined, or the values just seem to lead to social relationships that are anti-humanist. In Chapter 6, I then propose *power-based equality* as a necessary terminal value for meta-social good projects. In both capitalist as well as authoritarian socialist regimes, an accumulation of power leads to the degradation of institutions governing these regimes. An accumulation of power is also anti-humanist since it creates new social relationships that are coercive, as in the capitalist production process, or socially alienating, as with marketing products that have no real use-value for the consumers and do not lead to postive social relationships through production processes. Humanism is incompatible with the disempowerment of the weak or the instrumental use of others. Equality of power is also one of the key spirits behind democracy, so that each individual has an equal power to influence the outcomes of the democratic process, and those who are more powerful use their privilege to hear the concerns of the weak and empower them towards eventual equality. Meta-social good projects that aim for power-based equality can transform the systems of governance and the economy to prevent exploitation and dispossession of the poor, and provide more freedom to people to explore their opportunities and realize their capabilities. In Chapter 9, I then propose that a core instrumental value of *plurality*, which is a derived value from power-based equality, is needed for democratic processes to function and successfully find solutions that respect the diversity of the world. This diversity is essential to build a sustainable society that can be creative to solve the various problems that confront the world. I discuss that plurality in meta-social good projects of participatory societal communication systems that enable free expression, and facilitate the discovery of diverse views, are essential for a democracy to discover the essential values acceptable to society for social good. Further, Fuchs suggests that communication itself can be considered as a production process, and therefore should happen in ways that are not exploitative, coercive, or dominated by the powerful, which are ultimately anti-humanist (Fuchs, 2020).

It may appear as overkill to have society discover these values through democratic means that can lead to social good as humanism in production processes, instead of deriving them in some deductive manner and imposing them authoritatively on society through manifestos. As I will however discuss in Chapter 9, democracy is important not just as a mechanism to arrive at a consensus in a diverse society but also as a system that encourages people to understand diversity, reason and think critically on their own, and change their preferences through this process. Democratic systems are therefore essential for society to acquire the ability to learn, and I argue that core values of plurality and power-based equality are essential for this learning to happen freely and across differences.

Democracy is also essential to control technology, as echoed by David Collingridge in *The Social Control of Technology* (Collingridge, 1980). Collingridge highlights a dilemma that the outcomes of a technology may not become known during its early stages, and when they do become known by then the technology is already popular and entrenched, and too difficult to change or displace. Consumer awareness and technology regulations to control these outcomes can only be discovered and enforced through the means provided by democracy. Democracy is thus the vehicle through which society can impose its control on technology, and meta-social good projects that strengthen democracy are therefore essential to promote technologies aligned with social good and reject those that can be disempowering.

Fig. 3.1 summarizes the different elements of doing social good. Society creates and legitimizes the social systems to govern itself. Meta-social good projects ensure that these social systems do not create disempowerment and exploitation. They support society to build an informed public opinion, and the technologists within society to build a humanist ethos in their work, so that society can discover values that should be the basis for social good, and technologists can build systems that adhere to these values. In this way, meta-social good projects create a conducive environment for social good projects to flourish.

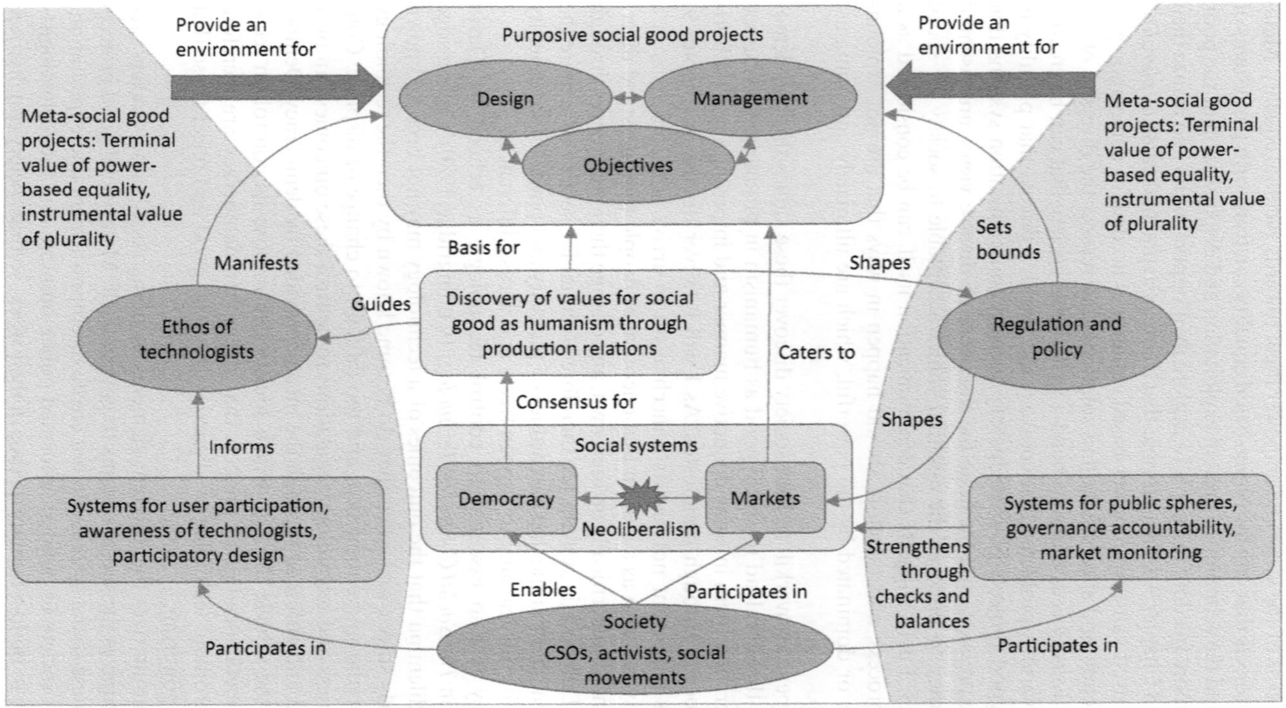

Fig. 3.1. System Map of Interactions Between Technologists, Society, Institutions, and Technology Projects for Social Good.

This view of enabling meta-social good systems that are centred on plurality and power-based equality, to allow society to think critically and to democratically discover values that define social good with a goal to increase humanism in society, is similar to the notion of a pluriverse introduced by Arturo Escobar in *Designs for the Pluriverse: Radical Interdependence, Autonomy, and the Making of Worlds* (Escobar, 2018). Escobar envisions a society that will, through autonomous bottom-up mechanisms, evolve its own design principles when it is free of any influence exercised through hierarchies of knowledge or discourse, and it embraces values of collectivism over individualism, matriarchal values (such as inclusion, participation, collaboration, understanding, and respect) over patriarchal values (such as competition, hierarchies, power, growth, domination of others, and appropriation of resources), a one-ness with nature over a dualist subject–object perspective of nature, and is skeptical about rationalist approaches that are justified as truth. Escobar sees such a pluriverse as an end-state for post-human societies where many worlds can flourish and creatively construct new designs for living. Although I do not go as far as Escobar, my conceptualization of meta-social good projects is similar in the sense of such projects as enabling an environment that is driven by plurality and power-based equality within which society can discover values that it deems necessary for social good. Further, equating social good with humanism constructed through social relationships of production is similar to values of collectivism and collaboration as suggested by Escobar. Establishing such an environment however requires a transformation of the current hegemonic systems of the world that are quite the opposite – hierarchical, exploitative, individualistic, rationalist, and modernist.

The need for system transformation as essential for genuine social good has lately also been considered in a recent theory on transformative social innovation (Haxeltine et al., 2016). Social innovations are suggested to be evaluated on dimensions of their nature of activities, values driving these activities, and power outcomes as a result of the activities. This is similar to the framework I have suggested, of expressing projects in terms of their terminal values to achieve certain outcomes, instrumental values that govern the activities of the project to achieve the desired outcomes, and to have power-based equality as a core terminal value and plurality as a core instrumental value. If power dynamics are not considered then the projects may even lead to the disempowerment of those they meant to empower (Avelino et al., 2017; Seth, 2020c). It is important to also note that executing social good projects, or the transformational meta-social good projects, will not be straightforward though, because of the hegemony of the capitalist ideology that suppresses anything that challenges it. In Chapters 7–9, I discuss strategies that technologists can adopt to challenge this hegemony.

To summarize, I have suggested in this chapter that social good is purposive and consequentialist in nature, and it should be specified in terms of ethics. Frameworks such as VSD or the capability approach are ideal to express ethics-based foundations for social good projects since they allow a distinction between terminal and instrumental values; participatory democratic means should be used by society to discover the terminal and instrumental values that it associates with social good; and it is feasible for all technologists irrespective of where they are situated in the production value chain to relate these underlying values with their work. Social good projects can then be identified as those that follow these values to increase humanism in society by creating positive social relationships through production processes, and a further category of meta-social good projects are needed to ensure that these values are adopted by apparatuses of the state and markets that are currently prevalent in the world to impose social control over technology. The terminal value of power-based equality and instrumental value of plurality can lead to a transformation of the prevalent systems of the state and markets to prevent the exploitation that is caused by them. Technologists can use this framework to reason whether the projects on which they are working are actually doing social good or not, and accordingly realign their priorities. As we have discussed earlier, humans derive their humanism from being engaged in social relationships of production to meet genuine use-values in society, and not being able to do so is ultimately alienating for them. Only by being engaged in social good projects that increase humanism in the world can technologists avoid this alienation. This is the new paradigm of technology development that can avoid disempowerment of the weak and support technologists in finding their humanism.

Chapter 4

Ethics-based Foundations

In the previous chapter, I demonstrated that social good projects can be described in terms of ethics-based foundations, with a clear notion of terminal values that the project aims to achieve, and instrumental values to which the project aims to adhere. In this Chapter, I describe a framework that can help determine whether a project is indeed complying with its specified or claimed ethics-based foundation by checking for internal consistency among the ethical values embedded in different design and management components of the project. This chapter also serves as a quick introduction to common fault-lines where ethical issues arise in technology projects, and current approaches to address these issues. More detailed material is available as part of an introductory course on computer ethics taught at IIT Delhi (Seth, 2020a).

A technology project can be conceptualized as being constituted of several design and management elements, and a model of how it intends to achieve the planned social good goals. Each of these constituent elements have ethics-based foundations of their own, and a consistency check can be performed to specifically evaluate whether they are indeed in agreement or in conflict with one another, and with the social good goal. This can help verify whether the means of achieving social good are in agreement with the ends. Projects that are not internally consistent in their ethical values can be ruled out as doing social good. This approach can also help make legible and bring transparency to the underlying ethics-based foundation of a project. This can assist its consumers or users to determine whether their own ethical preferences are aligned with those of the technology project, and thereby even choose whether or not to adopt a particular technology. Similarly, a clear articulation of this ethics-based formulation can potentially serve a statutory role as well of how a government, for instance, may determine the eligibility of a project to join a social stock exchange. In subsequent chapters, I further suggest some core ethical values that projects should follow to qualify as social or meta-social good projects.

Built through a detailed literature survey, I describe a framework of constituent design elements and management practices where ethical questions commonly arise in technology projects (Seth, 2020a, 2020e, 2021a). Fig. 4.1 shows this framework. The framework has been split into two layers: design and management.

Technology and (Dis)Empowerment: A Call to Technologists, 55–81

Published under exclusive license by Emerald Publishing Limited
doi:10.1108/978-1-80382-393-520221004

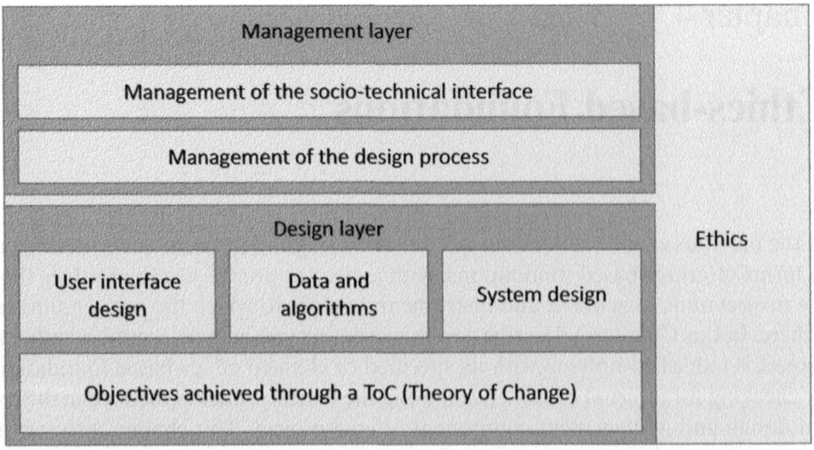

Fig. 4.1. Constituent Design Elements and Management Practices for Technologies.

The design layer in a technology for social good project can consist of several elements:

1. The Theory of Change (ToC) outlines how the technology project will achieve its stated social good goals (Clark & Taplin, 2012). As discussed in the previous section, social good is essentially consequentialist, similar to social development, and hence the ToC method widely used in the social development space can also be used to describe assumptions of how the technology is expected to achieve the desired social good goals. These assumptions can be validated against the ethics-based foundation of a project to check whether indeed they are in compliance with each other.
2. The user-interface design for technologies is relevant when they are meant to be used directly by end-users. Ethical questions arise often when designing user interfaces, as with having anthropomorphic elements or persuasive design techniques that can nudge users towards certain actions prescribed by designers without prior consultation or knowledge of the users (C. Schneider et al., 2018). These raise concerns about respect for human rationality and informed choice between the users and designers (Duquenoy & Thimbleby, 1999).
3. The methods for data collection, storage, and use in the project. Easier copying, search, and analysis of data that comes with increasing digitization has led to significant concerns with data privacy and the ethical use of data. These issues are now widely known and have moved far beyond being of just academic research interest, and several aspects are already enshrined in law in many countries (GDPR, 2016; Miller, 2013; Wire, 2016). Values shaping the data policies of a project can be checked against its ethics-based foundation.

4. Algorithms that operate on the data. Algorithms digitally encode objectives that are computed on tractable models, and this has raised questions about the fidelity of discrete categories that define the variables modelled in the algorithms (Crawford & D. Boyd, 2012; O'Neil, 2016), the ethics of the objectives themselves and biases built into them (Hindman et al., 2003), additional biases that may arise with data-driven algorithms and lead to discrimination against already marginalized social groups (Angwin et al., 2016), and the multiplicity of ethical definitions to codify these biases and guide computational methods to avoid unjust biases (Verma & Rubin, 2018). All these aspects can be linked with the underlying ethics-based foundation of the technology system.
5. The system design that outlines how technology intermediates to create and modify social relationships between different stakeholders that interact directly and indirectly with it. This I describe in more detail in Chapter 6 when I argue for power-based equality as an essential terminal value for social good projects (Seth, 2019a; Winner, 1980). This value can guide whether to set up projects that foster more equal power relationships, for example, through centralized decision-making elements versus a decentralized design, assisted access versus private access, or aimed at collective interactions versus individualized interactions. This element becomes more important especially in the context of technology platforms such as Facebook or Uber which are essentially sites that intermediate interactions between multiple direct and indirect users, and thereby shape the social relationships between these users through the codified usage and governance rules of the platforms (Geiger, 2015).

I consider all of these above elements as design elements. Of these, the ToC would be an essential part of technology for social good projects to define the pathways through which the project envisions to achieve its goals. The other design elements service these pathways. The ethics embedded in these elements serve to enable or constrain the affordances in how the technology projects are used so that they adhere to the pre-determined pathways while avoiding harm (L. Sanders, 2008; Vredenburg et al., 2002).

The notion of design that a blueprint comes first and it then shapes the usage of the designed objects accordingly, as most design methods are framed (Blomberg et al., 1993; Ideo, 2008; E. B. Sanders & Stappers, 2008; L. Sanders, 2008; Spinuzzi, 2005), is limiting though. This is simply because many concerns unforeseen at the design stage arise at the socio-technical interface of technology when it is deployed and used by people, post design. This may be due to a lack of adequately diverse prototyping or due to surprises that are bound to arise when technologies are deployed in a world that is immensely complex. Often dismissed as *unintended outcomes* and followed with debates about accountability, these issues however require careful management to address them, such as by re-designing the technology or improving governance of the technology usage. I assert that management practices to watch out for such emergent issues during deployment, and to handle them, should also be examined from the same ethics-based foundation of the technology project as what guides the design. This is why I indicate management practices as a separate layer in the framework shown in Fig. 4.1. Although most design literature often does not distinguish between design and management practices, considering management methods as a part of design as well (Escobar, 2018), I prefer to draw a distinction between the two because, in practice, the streams are operated differently by different teams, and the respective challenges within each stream are also different. Management practices need to astutely monitor and assess whether the envisioned objectives of the technology project are being realized, and provide feedback to the design layer through an iterative process so that the design can be suitably altered to bring the project back in line with the desired goals and ethics. A lack of attention to management at the socio-technical interface can be misleading – approaches of ethics by design alone can solve some problems but not all, and relying on them alone could in fact even give a false sense of safety.

This ethical framework can be used by technologists at three levels:

1. It can help examine any technology in terms of immediate concerns that might arise through its choice of design elements or management practices, and whether their ethics-based foundations are internally consistent with one another and with the project goals. This can equip technologists to course correct and minimize harm, and also quickly reject projects that exhibit an internal inconsistency in the values of their constituent elements.
2. The specific choices of terminal and instrumental values that are selected can be used to reason whether the technology can be considered as being meant for social good or not. As discussed earlier, which values are considered as being good for society should be decided by society and enacted through its institutions. Further, the presence of terminal values should be a necessary qualification for social good. The ethical framework can therefore be used to check whether indeed terminal values have been clearly specified, and whether the constellation of terminal and instrumental values are in agreement with those chosen by society for social good.
3. Technologies can be examined to understand whether they are aimed at social good or meta-social good. Technologies that facilitate society to democratically debate which values are necessary for social good, or technologies that strengthen institutions such as the state and markets by holding them accountable to society to uphold these values, or technologies that counter the dominant hegemony of capital to incorporate these new values in capitalism, are what I termed as meta-social good technologies. These aim to create systemic transformation to provide a conducive environment for social good projects. Technologists can thus use the framework proposed in this chapter to reason at which level are they operating, and accordingly align their priorities to simply minimize harm, or to go further to do social good, or even further to enable the conditions for doing social good.

4.1 Ethical Framework to Examine Technologies

I next outline each of the design and management elements in more detail, and describe methodologies through which their ethics can be discovered. I apply these methods in the next section on some examples of technology systems.

4.1.1 Design Layer: Theory of Change

The ToC of a project is a deterministic outline of how the inputs to a project are supposed to lead to the desired outcomes (Clark & Taplin, 2012). It is a disciplined method to describe the causality assumptions behind the expected changes. Often it is shown in three stages, of inputs, which lead to intermediate outputs, and which in turn lead to eventual outcomes (N. Hansen et al., 2019). The linkage between the successive stages is described through the causality assumptions. Inputs would consist of the technology system being put into place, and other elements such as any trainings to be imparted to different stakeholders, or marketing of the technology system to advocate for its adoption, staff recruitment, etc. Intermediate outputs are typically stated in terms of metrics such as the number of users, activity intensity of engagement on the technology system, and other usage milestones. Eventual outcomes are stated in terms of the actual change that results from the input activities and achieved intermediate outputs, such as changes in target SDG indicators (SDGs, 2015), or in case quantified development indicators are not suitable then in terms of qualitative changes such as achieved freedoms according to the capability approach (Kleine, 2010). The ToC therefore is a good starting point to outline all the design elements and management practices that are put into place, and how they are expected to contribute towards the desired social good goals.

It is important to clarify that having a ToC does not necessarily imply a top-down design that is imposed by external experts, which is the subject of criticism of the rationalist and inappropriate development methods conceptualized by the Global North and imposed on the Global South (Escobar, 1995). A ToC may rather be framed even in bottom-up ways in consultation with the communities for whom technology systems are being designed, and help logically define expected changes. Having a ToC that is legibly specified, or uncovered through interactions and discourse analysis with project participants, is useful to understand the underlying values that shape the assumptions behind the conceptualization of social good projects. Further, since social good projects are purposive, a ToC is important also to reason what should be the scope of project activities, and consequently to be able to attribute accountability for whatever outcomes arise from the projects.

Analysis method: An ethical examination of the ToC can be conducted to understand the underlying values behind the assumptions and envisioned pathways, and if the values are in line with the stated ethics-based foundation of the project. If a formally specified ToC is not publicly available then these underlying values can also be uncovered through interviews of the key project personnel and people impacted by the project. This can be done through methods like critical discourse analysis or most significant change techniques to examine the narrative stated by the project professionals, leadership, investors, users, and other stakeholders (Moitra et al., 2019; Wade, 2004).

4.1.2 Design Layer: User Interfaces

Nudging. Exploitation of known user biases can be used both maliciously as well as for good (Thaler & C. R. Sunstein, 2008). For example, the *Dark Patterns* website documents many cases where user interfaces insidiously shape choices of users to profit off them (Dark Patterns, 2020). This includes methods such as bundling of products on e-commerce websites that make cost-comparisons harder across different bundles, marking of default opt-ins to newsletters or purchase of value-added services (now deemed illegal under GDPR), introduction of hidden costs in the final stages of an online purchase, and putting up ads that are disguised as relevant clicks. More sophisticated nudging methods that ride on loss aversion biases of humans include the introduction of decoy options to persuade users to purchase higher valued products, and scarcity effects to persuade an immediate purchase by indicating that some options might soon become unavailable (C. Schneider et al., 2018). Other well-known biases include the anchor effect to show the least cost item first in a menu of choices, the primacy and recency effects where people commonly recall only the first and the last options, and the status quo bias where default selection of the existing choices makes it less likely for people to change their selections. Uber's user interface for drivers was found to contain such exploitative patterns to persuade drivers to stay on the road for longer, through the use of default selections to not go offline, and ludic-loop visualizations that put up income goals which appear to be tantalizingly close and nudge drivers to keep driving (Buchanan & Seshagiri, 2016).

Analysis method: Ethical theories like Kant's categorical imperatives to not use any human as a means to achieve your own ends would reject such a user-interface design (Quinn, 2014). The concept of nudges to influence behaviour for people's own good, in behaviourial economics, also emphasizes that the design should be transparent and never misleading, easy to opt out of, and be driven by strong evidence that the behaviour being encouraged will improve the welfare of those being nudged (Thaler & C. R. Sunstein, 2008). Such nudge patterns are well documented and can be spotted through a checklist-based auditing procedure to verify if a technology project contains such patterns, and whether they are consistent with the ethics-based foundation of the project.

Norm Shaping. Displaying metrics that can be perceived as a quantified measure of self-worth can lead to norm-shaping behaviour (Grosser, 2014). Social networking websites that show metrics such as the number of friends, likes, shares, etc., and allow benchmarking against similar metrics of others, leads to increased participation by the users in an attempt to improve their scores – a phenomena that has been described as *prescribed sociality for social performance*. Similarly, displaying time in terms of mere seconds since when new content was posted creates a norm for staying up to date – never to miss out on the most recent updates. Websites like the *Ledger of Harms* have documented many such patterns that are used to monopolize the attention of users, but which can also lead to mental health problems, confused human relationships, and learning and socialization issues among children (Ledger Of Harms, 2020).

Analysis method: Such design methods raise concerns of individual autonomy and sociological and psychological issues that can arise from norm-shaping design. Auditing procedures can help check for the presence of such norm-shaping patterns in technologies and qualitative studies with users to understand their experiences with and without metrics can help ascertain if the metrics are indeed shaping people's norms, and whether the norms are in line with the stated social good goals (Grosser, 2014).

Anthropomorphic Properties. Psychologists have observed that people attach a personality to computers that can be manipulated through known biases, such as the use of bold fonts or authoritative statements in computer interfaces convey a dominant personality, and when computers use colours or logos liked by a user then their recommendations are considered more agreeable (Fogg, 2003, 2009). Personas coded into computers to evoke feelings of happiness, anger, fear, helpfulness, praise and flattery, goofiness, etc., can become a means to mislead users into actions that they otherwise may not have taken in the absence of the machine. This raises concerns on autonomy of the people, and of complexities in placing accountability on machines for the outcomes influenced by the machine.

Attributing accountability can become more complex to understand with advancements in artificial intelligence (AI) and robotics where the machines themselves make decisions driven by complex algorithms (Wiener, 1950). Relating a personality with machines that are driven by algorithms, especially data-driven algorithms that are not easily explainable, raises additional concerns about how human behaviour may itself get influenced when they interact with machines that can operate much faster than what humans can keep up with or understand (Allen et al., 2006; Moor, 2006). These dilemmas led Joseph Weizambaum to warn against building anthropomorphic technology in the first place since it could diminish people's respect in human agency, especially if a human function like love or interpersonal respect were to be substituted with a computer system (Weizenbaum, 1976).

Analysis method: Contrary to Weizambaum's advice, chatbots and virtual agents are becoming popular in giving support to people to discuss personal issues including about mental health, and are in fact positioned as a use of technology for social good (Følstad et al., 2018). It therefore becomes important to clarify the ethical foundation of such projects. The projects may use consequentialist arguments that the users ultimately benefit from the interactions, and augment them with deontological arguments to justify user autonomy since informed consent procedures are in place. However, to judge the credence of these arguments, it should be mandatory to hold clinical trials and incorporate learning from other projects where the implications of anthropomorphic design may have been evaluated. This requires values of honesty and thoroughness in research about these technologies.

4.1.3 Design Layer: Data and Privacy

Among all elements of the design layer, data and privacy have probably received the most attention lately and have been theorized in multiple ways (Allmer, 2011; Tavani, 2007). The most common is a rights-based view of individual autonomy in the person having control over what personal information of theirs is shared with whom, how can it be used, and who else can access their information. This right to individual autonomy comes in conflict with collective goals of public safety though, such as when the state is allowed access to personal information of individuals through surveillance processes to impose control over them or others for the common good. Proportionality tests laid down in law are meant to decide and reason about the scope for data access in such situations, but they are effective only if a strong rule of law is in place to maintain data access within the decided bounds (IIF, 2020; Khan & P. Roy, 2019), and if technologies can be designed to provably adhere to the agreed upon data processing tasks only (S. Banerjee & Sagar, 2021). Regulations such as GDPR which do specify data processing principles like data minimalism and purpose specification are hard to enforce in the absence of clearly defined protocols to verify the data treatment claims made by the data controlling and processing organizations (GDPR, 2016). In web browsing, for example, methods like device fingerprinting and cookie synchronization to track users, and leakage of this data to other organizations, continue to proliferate because they are hard to detect and prove (Papadogiannakis et al., 2021). Similarly, technical methods such as homomorphic encryption or differential privacy which can restrict data processing to certain types of analysis, need wider adoption and even regulatory mandate in certain domains for such data protection laws to be effective (Agrawal et al., 2017).

Thinking of privacy only in terms of rules to control and restrict access to information about a person is limited in practice for another reason. The rules may need to be changed over time and customized to the situation in which the information is to be used, the context of prevailing social norms, and the personal preferences of people about maintaining bounds (Auston, 2019). Proactive steps therefore become important, such as to put into place programmes for privacy-literacy so that people can adopt appropriate privacy configurations as per their context, and to enforce data minimization principles through responsible user-interface design practices that make it harder for corporations to coax users to divulge more information about themselves.

Another view that counters the arguments favouring surveillance for security and social control discusses the relevance of individual privacy even for the common good because privacy is instrumental for diversity which in turn leads to innovation to move societies forward (J. Cohen, 2013). Further, surveillance is harmful for democratic societies because of its potential to quash dissent (Stallman, 2013). This viewpoint emphasizes that even the possibility of being under surveillance, that is, observability alone, is enough to curtail free thought which is essential for innovation required to solve complex societal problems (Flyverbom et al., 2015). Such behaviour modulation can thus be considered as individual privacy invasion, as well as being harmful for the common good of society by making society less agile to evolve and adapt to new challenges. The proposed data protection laws in India have been criticized as being too permissive in allowing government departments to access personal and non-personal data of individuals (Wire, 2016), and mark a significant departure from the desirable trend that had begun to emerge after the Snowden revelations in the United States for government departments to seek legal approval to access data of individuals (Miller, 2013). The mandatory use of Aadhaar for several government-citizen functions makes data integration even easier and susceptible to misuse or oversight as it now commands gate-keeping power for access to many essential services for citizens (Appadurai, 2020).

Zuboff argues through yet another direction of how surrendering private data to corporations that are interested in behaviour prediction and manipulation leads to a surrender of an individual's right over their own future (Zuboff, 2018). Further, the entire apparatus to do this was mobilized insidiously by corporations such as Google and Facebook without the consent of people. Through a process of incursion into private spaces without their permission, followed by habituating people to these privacy-violating technological tools, by the time these acts of violations were identified and understood to put up proactive regulations, the corporations had successfully learned to adapt to changing policies, even shape them in partnership with the state, and deflect demands for more effective regulation by ironically positioning themselves as the saviours of free speech. With digitalization leading to the availability of data that can be easily stored, analyzed, and transmitted, and the lag in regulatory oversight to control these processes, there has been a rapid acceleration in the erosion of privacy.

This use of data for commercial gain without having acquired consent has now become a key feature of digital capitalism for the financial gain of companies. This can be considered as a violation of human dignity and individual autonomy, where personal data is used as a means to shape people's purchase behaviour in ways that can harm them (Galic et al., 2017). Targeted advertising can even harm the common good by nudging people towards consumerism leading to a consequent negative impact on environmental sustainability (Nadler & McGuigan, 2018). Yet the convenience benefits of personalization have muted the criticism against it to some extent, even while information personalization has impacted democracy through targeted misinformation and selective information exposure (C. Sunstein, 2018; Wu, 2018). The notice and choice method of getting informed consent to access and use personal data is known to be inadequate because of the limited capability of any individual to understand complex data flows and how they may be put into use (S. Barocas & Nissenbaum, 2014). Capacity building of the users to understand these mechanisms therefore becomes essential if the procedures are to have any salience, as well as simplification of the complex legal language used in consent agreements so that people can understand them easily (Athey et al., 2017; New York Times Editorial, 2021).

Digital platforms have also failed in the digitization of complex privacy boundaries that people follow in real life in their interactions with others (J. Cohen, 2013). Consequently, privacy enhancing technological solutions are insufficient. Straightforward anomymization of data, for example, is unable to address problems where personal characteristics of many people can be revealed based on data about just a few (S. Barocas & Nissenbaum, 2014). The escalating need for solutions like *k*-anonymity and differential privacy have raised further questions about trying to solve problems for data that should not even exist in the first place (Agrawal et al., 2017; O'Neil, 2016). This impasse where unregulated and naively ambitious technological expansion has led to social problems that can no longer be addressed through technology alone, goes back to arguments by Weizenbaum (1976) about the need to reflect whether a particular technology is needed at all, rather than focus on how to fix the technology to make it more ethical. It underscores the need to have clear terminal values declared openly before new technologies are created, and to constantly use these values to guide the management of technologies as their adoption scales and deepens.

Analysis method: Data privacy can be examined against many ethical principles such as those related to individual rights, collective good, fiduciary responsibilities, interpersonal respect, and honesty of purpose. Examining the fine print alone of privacy policies of technology projects may not give enough information about what values are effectively realized, but being alert to news about privacy violations and data leaks can give hints about safeguards that may or may not have been enforced, and can reveal the underlying ethics of data protection followed in the project. Verified crowd-sourced registries such as the AI Incidence Database can be useful for this purpose (AIID, 2020). Blackbox auditing methods to check for evidence of information leakage can also be useful means to verify adherence to declared data protection policies (Diakopoulos, 2017).

4.1.4 Design Layer: Algorithms

Datafication. Simply speaking, algorithms relate input variables with a target variable, but deeper questions arise about why a particular target variable or objective was chosen, how the input variables were chosen, and how the variables were discretized (Gillespie, 2014). Discretization into well-defined categories can be restrictive, as was understood recently in the context of gender categories on Facebook, and has been known more generally with the challenges of digital and analogue representation (Lees, 2014; Wilder, 1997). Algorithms that model humans to help make decisions on their behalf, are only able to model a shadow of the person as required for the purposes of the objectives coded in the algorithm and to make the computation tractable (Gillespie, 2013). These models have been referred to as the *algorithmic self*, in contrast to the *actual self*. Similarly in the context of AI, the simplified representation of the world that machines are able to process has been stated by Luciano Floridi as *AI inscribing the world* to make it legible to machines (Floridi, 2018). The success of informatization of natural processes like protein interactions, or industrial automation that relies on physics modelled precisely in engineering systems, does not easily generalize to modelling of social processes. For example, this simplification is known to often lead to mistaken labelling of the social group to which people may belong or identify with, and may be done without their knowledge, which can have discriminatory outcomes (Mantelero, 2016; Vries, 2010).

Despite this subjectivity in making choices about how different variables are expressed, which ones are included or excluded from a model, and ethical underpinnings behind these choices, algorithms are still perceived as authoritative by many people. Rather than building interfaces for humans to guide machines, AI is often presented as doing tasks better than humans. This can lead to the pursuit of automation or datafication and digitization as an end in itself, and diminish the faith that humans place in their own agency (Scannell, 2015). This may effectively even create a new *knowledge logic* arising from a mistaken faith placed in knowledge created by algorithmic models, instead of knowledge produced by credentialized experts or the scientific method or even common sense (Gillespie, 2013). Considering the algorithm as an independent entity, separate from its creator or administrator, also impacts accountability by effectively creating a *moral buffer* that distances the technology creators from liability (Gillespie, 2014). All of these raise ethical concerns, as described next.

Algorithmic Objectives. I consider the example of web search to outline ethical questions behind the choice of objectives that are coded in web search algorithms. Web search has become essential to use the Internet for access to information. The term *Googlearchy* was coined almost two decades back to denote feedback loops that web search algorithms using page-rank create. Page-rank ranks higher those pages that have more incoming links from highly ranked pages. This leads to already popular pages being ranked higher and therefore having a greater chance of acquiring more links to become even more popular – a clear *rich get richer* phenomenon created by the specific choice of algorithmic objective (Hindman et al., 2003). Other sources of bias like self-selection by biased people in creating links aligned with their bias, or biases due to an inherent homophily in social networks of users through which information is filtered, further influence the ranking of content (L. A. Adamic & Glance, 2005; Bakshy et al., 2015). This biased ranking can shape perceptions of people, or even radicalize users or harden their confirmation biases (Epstein & Robertson, 2015; Roose, 2019). This can not only lead to grave individual harm by influencing personal behaviour, but also collective harm by impacting democracy (C. Sunstein, 2018).

Another example where the choice of algorithmic objectives raises ethical questions is with algorithms for resource allocation, such as welfare benefits (Aiken et al., 2021; M. Lee et al., 2018). Here, a choice can exist between having a short-term allocative metric such as *equality* coded in the algorithmic objective to distribute scarce resources, or a long-term allocative metric such as *equity* to make future allocations contingent upon past unfairness in distribution, or to consider different algorithms altogether that do not focus on making resource allocation but on removing factors that caused unjust resource distribution to take place in the first place (*justice*) (I. M. Young, 2011). This demands a clear ethical stance to decide which objective to use.

Biases in Data-driven Algorithms. Machine learning algorithms that learn based on observational data additionally suffer from biases arising from the data itself. Sampling bias occurs when inadequately representative data collection may have resulted in training the algorithms on only some underlying classes and not others (Buolamwini & Gebru, 2018). The classifiers thus trained would typically perform better on these well-represented classes and make mistakes on other classes. For example, face recognition algorithms make more mistakes on faces of black people than white people because the training data contained more samples of white people's faces. Another type of bias, called systematic bias, occurs when some spurious variables are highly correlated with the output variable because of some systematic noise in the data collection process (Ribeiro et al., 2016). Yet another type of bias called stereotype bias occurs when the training data contains a larger proportion of stereotype patterns, because the data collection process itself was biased towards these stereotypes. The popularly cited COMPAS study on recidivism in an example demonstrating this bias, where higher recidivism risk-scores were predicted for black people than for white people (Angwin et al., 2016). This was because blacks were more actively policed than whites and consequently re-arrests of black people were more common than that of white people, hence machine learning classifiers ended up learning a relationship between race and re-arrests, using re-arrests as a proxy for recidivism. Entrenched societal discrimination reflected in the training data was therefore reproduced by the classifiers trained on this data. Blinding the classifiers to remove the race variable to prevent disparate treatment on protected variables is typically not sufficient because similar patterns end up getting inferred through other correlated variables like education or income. Several alternative fairness definitions have been proposed to counter such biases at the outcome stage to prevent disparate impact, but different definitions were found to be more preferable by different stakeholders: defendants would want false positive rates to be low to minimize the chances that they are incorrectly classified as high-risk, whereas decision-makers would want the precision to be high to ensure that those labelled high-risk are indeed those who recividate (A. Narayanan, 2018). Impossibility theorems further show that satisfying several definitions simultaneously is not possible, and therefore technology designers have to favour one or the other stakeholder in choosing a fairness definition to use in the algorithmic decision-making technology (Chouldechova & Roth, 2018). The book *Fairness and Machine Learning* is an excellent reference to understand these aspects better (Barocas et al., 2019).

Scaffolding the Algorithms. Several research directions have also emerged to improve the acceptance of algorithmic decision-making and to ethically handle situations when algorithms fail or are biased. Explainability especially of data-driven algorithms has been recognized as being important to users to understand algorithmic decisions (Eslami et al., 2018; Gilpin et al., 2018). This can also help dispel misperceptions about the authoritativeness of algorithms by exposing their fallibility when their explanations do not make sense. In cases of incorrect decisions, accountability and appeal procedures need to be established through which people can question the decisions (K. Vaccaro & Karahalios, 2018). Some of these elements have been incorporated in regulations, although ambiguity remains about their implementation and practicality, as well as fears that lobbying and resistance by corporations for ease of business and economic growth can blunt them (GDPR, 2016; Larus & Hankin, 2021). Finally, human-in-the-loop methods are often used to leverage the algorithmic output as an input factor for humans to make decisions, rather than give complete control to the algorithms for decision-making (Green & Chen, 2019).

Analysis method: The complexity involved in the selection of input and target variables, their quantification, the formulation of algorithmic objectives, and fairness definitions, brings out the need for careful and informed decision-making when designing algorithms. An ethical evaluation of algorithms will need to check if these aspects have been reasonably considered. The mathematically grounded formal reasoning coded in algorithms can make legible their underlying ethics-based foundation (Abebe et al., 2020; Simons, 2019). Similarly, revealing details in privacy sensitive ways about data selection and collection for machine learning algorithms can help reveal different kinds of biases that may exist in the data. Errors still likely to persist in algorithmic decision-making may require providing appeal procedures and explainability to understand their cause. Skipping any of these steps can raise ethical concerns about honesty and attention to detail by technology providers.

Making algorithms and their underlying data or metadata transparently available can also allow for external auditing. In case, where full transparency cannot be enforced, with commercial technology projects, blackbox algorithmic auditing can be used (Barocas et al., 2019; Diakopoulos, 2017). These methods help reveal hidden biases by observing the output of the algorithms against curated test cases to check specifically for these biases. Such data-driven methods can thus verify if the values coded in the algorithms are in line with the ethics-based foundation of the project.

4.1.5 Design Layer: System Design

By system design I mean the concept of systems thinking and cybernetics that advocates modelling all stakeholders, and feedback loops between them, which influence the stability and evolution of the system (Beer, 1975; Meadows, 2008; Wiener, 1950). The view that technology can influence social relationships between different stakeholders that engage with and use it, directly or indirectly, has not received a lot of attention in the design literature but this view can be useful to build a comprehensive understanding of how the technology may affect the broader ecosystem in which it is embedded. I discuss this more in Chapter 6 and present an overview here.

This notion that a technology can influence social relationships between different stakeholders was prominently highlighted by Langdon Winner through several examples (Winner, 1980). One such example was the creation of low-elevation road bridges that prevented buses from entering certain neighbourhoods, and thereby ended up restricting entry for low-income populations (often blacks) who travelled using public transport, while elite groups with personal cars were able to travel freely. Such concerns were in mind with early work in participatory design where information technologies introduced at the workplace were examined critically from the view of how they would influence power relationships between the management and workers (Spinuzzi, 2005). Similar issues were considered in the ETHICS methodology (Mumford & Weir, 1979). However, design thinking since then evolved more in the direction of designing technology for individuals (Ideo, 2008) or for collaborative work in groups (Poltrock & Grudin, 1994), or examining from an interpretive rather than an interventionist or designer viewpoint about the role played by technologies in shaping social systems (B. Latour, 1996). Counter examples are few, like designing technologies aimed at fostering trust between citizens and civic agencies, or technology that can balance power relationships between beneficiaries and officials for inclusive access to social protection (Harding et al., 2015; Seth, 2020c). Such interventionist approaches to shape social relationships are otherwise seen more commonly in the social development space, as part of articulating the ToC of a programme, such as to improve gender equality within families and communities or to strengthen the accountability of governments towards citizens.

System design setup and goals are especially important to consider in digital platforms since these platforms define how different stakeholders engage with one another and thereby how their social relationships are shaped. For example, Uber uses the ratings given by customers to drivers to decide how to allocate future rides to the drivers; this creates imbalanced power relationships between the customers and drivers (Rosenblat & Stark, 2016). In another example, a Management Information System was set up for an employment guarantee scheme in India to bring transparency, but actually enabled senior bureaucrats to monitor the activities of junior bureaucrats, and who in turn developed counter tactics through mis-reporting of data to continue with corrupt practices at the last mile (Veeraraghavan, 2013). Similarly, the introduction of Aadhaar-based biometric authentication to access welfare benefits was envisioned to take discretionary power away from local government officials and politicians, but introduced a new set of intermediaries to enable people to interface with the technology (R. Khera, 2017).

Such rules governing technology platforms are built into the code of how the technology of these systems operate, and can be considered as a form of governmentality not very different from bureaucratic setups where officials follow procedures to govern markets, welfare systems, etc (Geiger, 2015; Lessig, 1999). Human-in-the-loop systems attempt to take away some extent of governmentality from the technology and leave room for human discretion, but this may increase the chances of emergence of imbalanced power relationships between users and the bureaucrats operating the technology system (Bullock, 2019). Such aspects require careful ethical consideration of the power held by designers and administrators of the system. Some systems may try to involve users in a democratic governance model for rule-making, but need to be examined with regard to the extent to which user participation is entertained by the technology providers (Martin et al., 2017).

Analysis method: The system design for technology projects can raise many ethical concerns, based on whether participation in the system affects different stakeholders differently, or whether some design elements are discriminatory towards a particular group of people, or the underlying assumptions themselves can have their own ethical basis favouring competitive versus cooperative behaviour or centralized versus decentralized control, etc. An ethical analysis of the system design may best be revealed through surveys and field studies that identify direct and indirect stakeholders impacted by the technology, and understand their experiences before and after the introduction of technology in terms of how it altered their social relationships with one another (D. Chakraborty, 2018; Seth, 2020b; Veeraraghavan, 2013). The modeling approach described in Chapter 6 for how power relationships between these stakeholders are altered by technology projects, can help determine who gets empowered and who does not. Critical theory using subaltern[1] approaches to understand voices of marginalized stakeholders can also be essential to help reveal if the system design is indeed empowering for them (Masiero, 2018).

4.1.6 Management Layer: Managing Design

The design elements described so far are created through a design process. I consider the process of coming up with a ToC (Theory of Change) and the technology design, including their ongoing evolution, as part of the management layer – a sort of meta-design to govern the design process. The ethics of this design management should also be compliant with the underlying ethics-based foundation.

Designers typically hold more power than users, and different underlying values of the designers like empathy, sympathy, or apathy, can lead to different design outcomes (Duquenoy & Thimbleby, 1999). Empathy is stated as the only value that would help the designer see the user as the end, and not a means for other objectives (Vistisen & Jensen, 2013). Anything other than empathy could become autocratic by bringing in biases and agendas of the designers (Vandenberghe & Slegers, 2016). It is suggested that user engagement and their participation in the design process should be a mean to understand the users and accordingly respond to their needs or aspirations (Irani et al., 2010; K. Toyama, 2017). This comes back to participatory design and the method's commitment to building user capacity so that they can effectively influence the design (Xiao et al., 2004). Participation may not always be easy to incorporate though. How participation at the design stage is configured in the project in terms of representation of different stakeholders, can be influenced by practical aspects. These could include the amount of time that prospective users are able to dedicate towards the participatory design process, language differences between the users and designers, how design meetings are facilitated, and comfort between the users and designers to speak freely (Heeks, 1999; Pedersen, 2015; Vines et al., 2013). The workers' inquiry method (see Box in Chapter 7) from a Marxist inspiration acknowledges such aspects and recommends discussions and reflection over an extended period of time among all participants including the workers themselves to truly understand their own context and intervene accordingly to overcome oppressive power structures (Haider & Mohandesi, 2013). Participation may however also end up being just a facade though – designers may not truly have genuine intentions to use the insights generated through the participatory process (Arnstein, 1969).

[1]The *subaltern* refers to subordinated populations that are systematically objectified by those in power, which muddles their true voice and effectively de-voices them (Spivak, 1988).

Analysis method: The diversity of design methods has been discussed as being analogous to the diversity of ethics itself, and should be compared against the underlying ethics-based foundation claimed for the technology project (Cairns & Thimbleby, 2003). Values such as human dignity, equality of participation, taking responsibility, and continuous evaluation, can be important considerations for an ethical management of the design process. To understand the ethics of this process, maintaining an ongoing documentation log of all design changes, discussions with different stakeholders, and decisions made, can provide a historical view of the design process (Neyland, 2016).

4.1.7 Management Layer: Socio-technical Interface

Once technologies are designed and deployed, many new concerns can arise at the socio-technical interface when different stakeholders begin to use the technology (Seth, 2020e). This can especially happen when certain users engage in improper or unexpected use of the technology, which 'makes the idea of total design control problematic' and requires action from other stakeholders such as managers or external institutions to restore the system to its intended functions (Kroes et al., 2006). Appropriation of social media platforms for disinformation or hate speech is a prominent example (Arun, 2019). Management of these emergent concerns requires an ethical approach as well, and to evaluate it against the stated ethics-based foundation of the project.

Such concerns could be about what policies should be developed to shape appropriate usage norms of the technologies (Chandrasekharan, Samory et al., 2018; Lampe & Resnik, 2004), who ends up being included or excluded from access to the technologies (Barboni et al., 2018; R. Khera, 2017), whether accountability mechanisms are in place to have different stakeholders engage with one another in mutually respectable ways (K. Vaccaro & Karahalios, 2018), and how even changes in goalposts or the ToC may be accommodated in response to changing social and economic conditions (Moitra et al., 2016; Seth, 2020d). These concerns can be managed by putting specific processes in place, which would have their own ethics-based foundations as well.

Inclusion and Exclusion. Similar to the project scoping considerations in the market prices example discussed in Chapter 3, several management practices can be important to keep a check on inclusion and exclusion biases that may arise in technology access (Seth, 2020e). Tracking of demographics of users who access the systems can give insights on who is benefitting and who has been left out. Financing of access for excluded users can help ensure equality in access. Capacity building based on proportional assistance given to vulnerable groups can improve equity of outcomes realized from use of the technologies. Technology providers can then be evaluated on the extent to which they take accountability and respond to emergent gaps in meeting the project goals. Bringing continuous feedback through user surveys, focus group discussions, user interviews, and grievance redressal channels through which users can raise issues and give feedback, can also be important to understand various concerns and address them.

Controlling Usage. The example of content moderation on participatory media platforms is a rich source of learning where many mechanisms have been used to control and shape technology usage. Managing concerns such as misuse and misinformation, censorship, manipulation and bias, management of discrimination, and abuse of market power, are some aspects that needed careful platform governance (Saurwein et al., 2015). Some platforms have adopted a market-based approach to governance that effectively outsources the responsibility of respectful usage to letting the users develop their own norms. In such platforms, user-facing tools are used to flag objectionable posts and involve users in the moderation process (Crawford & Gillespie, 2016). Seemingly simple decisions like whether or not to expose the moderation history publicly (as in Wikipedia) to highlight underlying conflicts about the agonistic character of moderation, can have wide implications to encourage values like deliberation and diversity among the platform users (Crawford, 2016). Users opposed to diversity can sometimes resort to uncivil dialogue, and signalling mechanisms like banning such users or removing their comments can further convey a commitment to uphold values of pluralism and diversity (Chandrasekharan, Pavalanathan et al., 2018). Effective redress mechanisms available to users to appeal against governance decisions are also an important element to ensure fairness and justice to users. When regulatory measures like censorship are invoked by platform providers to determine what is permissible or not, concerns also arise about how such regulatory standards were developed in the first place. Facebook for example continuously evolves its community standards but it is not transparent about the process through which the standards are developed and maintained. Even governments have lately intervened to influence speech on these platforms (Srivas, 2021). Alternate governance structures like federated platforms, where different groups are empowered to institutionalize their own regulatory methods, opens up the wide diversity in platform governance models that are seen in practice (Chandrasekharan, Samory et al., 2018; Seth, 2020e).

Such variants of platform governance emerge from different ethical positions and can be examined to check if they are in agreement with the stated ethics-based foundation of the technology system. Values about promoting mutual respect among users, pluralism and diversity, capacity building of users, and empowering users to influence the working of these technologies in a democratic governance model, can appear in sharp contrast with non-transparent control mechanisms that are in use by Facebook, for instance. The examples discussed later in this chapter highlight such conflicting ethical perspectives that are seen in practice in the governance of many technology platforms.

Experimentation for Better Management. Many technology platforms actively run experiments on their users to test new features and design changes. The infamous emotional contagion study by Facebook on 700,000 users, without the users knowing that they were the subjects of these experiments, revealed how research ethics procedures were circumvented by claiming the experiment to be a quality improvement exercise (Grimmelmann, 2015). Such experiments that are deployed with the intention to test new ways to regulate user behaviour, raise additional concerns on user autonomy and social engineering (Popper, 1945). In response, community-centric methods have been proposed that respect values of user consent and deliberation to first agree upon the cost-benefits of running population scale experiments and then rolling them out only upon achieving consensus (Matias & Mou, 2018).

Concerns also need to be addressed about the scientific validity of running such experiments. Measurement challenges may arise from sampling biases based on who opts to participate in the experiment, or the *Hawthorne effect* when users may be aware that they are being observed (Campbell, 1991). When studies are done on public technology platforms for research then concerns may also arise on the openness of making data available for other scientific uses.

Evolving Management. My goal with distinguishing management from design was specifically to highlight that not everything can be planned in advance. New challenges will inevitably emerge and will require new ways to deal with them. Especially when attempting to do social good, Rittel and Webber characterized many social problems as wicked problems to which no clear solution is possible - they are only to be *re-solved over and over again* (Rittel & Webber, 1973). Therefore even the approach to management will need to learn from experience and evolve. It is entirely feasible that the initial ethics-based foundation laid down for the project, may also evolve with new values recognized as being important, and old values being deprecated or deconstructed upon deeper reflection. How can technology management itself be guided to evolve and grow?

One method is by spotting conflicts that may arise between users, or grievances raised by users to the platform managers. Conflicts can reveal ethical positions of technology systems, and also shape their further evolution (Bodker & Kyng, 2018). Protests by users on Reddit against the firing of a loved Reddit employee, for example, revealed the strong sense of ownership by Reddit users to exercise their voice, and led to the platform providers conceding more flexibility to the users (Centivany & Glushko, 2015). Such genuine participation by different stakeholders in the management process is also useful to keep the technology providers informed about users' concerns and navigate the way forward accordingly (Arnstein, 1969).

Some industry centric methods have also been developed. Software Development Impact Statements (SoDIS) relies on risk-assessment and Responsible Research and Innovation (RRI) relies on self-reflection for organizations to guide their evolution and decision-making (Gotterbarn & Rogerson, 2005; Martinuzzi et al., 2018; Rogerson, 2017). Sincerely following such processes can highlight the ethical commitment of technology providers towards honesty and welfare of their users.

Analysis method: To understand the underlying ethics of the management processes, methods such as the following can be used. Discussions with different stakeholders and maintaining an ongoing documentation log of concerns that arose at the socio-technical interface, how were these concerns addressed, and any changes consequently made in the management processes, can provide a historical view of the development of these processes to understand their underlying ethics (Neyland, 2016).

4.1.8 Other Frameworks for Technology Ethics

In contemporary debates about ethics, especially with the emergence of AI, its role has predominantly remained restricted to technology design (Greene et al., 2019; Washington & Kuo, 2020). I have attempted to show that the scope of ethics goes beyond design, to management of the design processes, and processes for management of the socio-technical interface of technology. Similar views were expressed by Hans Jonas in discussing the uncertainty with many new technological innovations in their influence on future generations, and hence he too emphasized that usage principles should evolve continuously through oversight and monitoring as new concerns and conflicts arise when technology is deployed (Jonas, 1985). Consequently, the ethics-based foundations of projects may need to evolve as well – ethics can therefore neither be considered as static nor as just confined to design (Frauenberger et al., 2017). This aspect has largely not been considered by design manifestos (Mulvenna et al., 2017), nor by design methods like VSD (Friedman et al., 2013), or computational methods to address biases (Chouldechova & Roth, 2018), or mechanism designs for deterministic outcomes to emerge from technological systems (Cantillon, 2017). This shortcoming of design methods has also been inherited by policy-makers and governments in their deterministic thinking about technology (ACM, 2017, 2020; IEEE, 2014).

Probably the closest to having a comprehensive method is *action research*, with its emphasis on several key aspects that have been discussed above: the participation of the community, securing agreement from them for any intervention, joint development and ownership of the interventions, and careful monitoring and re-design (G. Hayes, 2011). Action research appears too utopian, though, to expect that technology companies will adopt these methods in a fiercely competitive landscape, where speed of execution and quick traction determines access to financial capital. The method I have proposed might be more practical, to continuously examine what values seem to be in play to explain the management steps taken and to compare them against a stated ethics-based foundation. Inconsistencies can be flagged for corrective action, or discussed publicly to draw consumer attention to these concerns.

The framework proposed here can be used to understand the design elements and management practices in a technology system. I have also described in brief various methodologies that can help uncover the underlying ethics-based foundation of these elements and relate them with the stated positions of the technology project. Similar frameworks have been proposed by others. Jacques Ellul proposed a list of 78 questions which span a similar array of design elements and management practices, and means and ends, to evaluate the ethics of a technology (P2P Foundation, 2019). Bernd Carsten Stahl and colleagues discuss how the underlying ethics-based foundations of emerging ICTs can be understood and addressed during the initial stages of the technology (Martin et al., 2010). They suggest similar methods like user surveys and interviews of project stakeholders to understand the ethical backdrop, and relate it to ethical theories.

Katie Shilton et al. suggest a framework that lists different kinds of values (social, instrumental, and environmental) and evaluates the extent to which a technology system adheres to these values (Katie Shilton et al., 2013). They differentiate between three types of adherence: salience, intention, and enactment. Salience is the extent to which a value is central or peripheral to the design elements and management practices. Intention is the extent to which a value is accidentally or purposively a part of the design and management of the technology system. Enactment is the extent to which a value has the potential to be exercised in the technology system. A democratically governed sharing economy platform is analyzed using this framework to determine the opportunity users have to influence the governance of the platform (Martin et al., 2017).

I bring forward some relevant aspects from these frameworks as I turn next to examine three technology systems.

Ethics of Informatization

Throughout this book, I consider a technology project as the unit of analysis, and examine it from various standpoints – the nuts and bolts of the design and management elements that constitute the project, the social relationships that it shapes, the political economy that shapes it, and the technology teams that govern it. Here, I briefly discuss computer technologies in general, abstracted away from specific projects in which they may be used, to see if there is a common way in which we can understand the implications of computers.

One way to view computers is that they take digitized representations of the world as an input, algorithmically manipulate these representations at high speed according to clearly specified instructions, and arrive at digitized outputs that are meant to serve as instructions to affect some changes in the world. This process termed as *informatization* requires therefore the datafication of any phenomenon so that it can be represented in a computer (Agre, 1997; M. Hardt, 2010; F. Webster, 1995).

This has been discussed in the context of industrial automation through computers during the latter half of the twentieth century that amounted to informatization of skills and knowledge of domain experts such as lathe machine operators who could then be replaced by computers. This shifted power away from them to programmers, and additionally created a skills polarization with white-collared programmers on one side and less-skilled machine operators on the other side (Braverman, 1974; Nobel, 2011). Machine learning appears to have made this process simpler where the task of formal representation is left to be discovered automatically by machines by identifying patterns in the underlying data. In a capitalist economy, information therefore becomes a commodity itself, be it in terms of raw data or derived patterns or programmed processes, and is valued based on its economic function or changes in power relationships affected by it, ultimately to extract more surplus value from labour (F. Webster, 1995). This is true also within industries that support manufacturing industries, such as advertising. Information is valued based on the predictability it introduces in consumer behaviour, and consequent control over consumer behaviour that it can facilitate.

Beyond informatizing domain experts or consumers, the same techniques are also used to informatize natural processes (F. Webster, 1995). Modernity, or the tendency of humans to control nature, has certainly seen many successes but the failure to account for externalities shows that informatization has happened selectively – the economics of information in the capitalist economy depends on what information is valued by the market, which is susceptible to many distortions as discussed earlier. The value of information is attached to production processes where it is used, and thereby depends on the market's recognition of the economic value of these processes. Some important processes may not be informatized, whereas other unimportant or undesirable processes may be informatized, based on misalignment between the market's beliefs and actual social good.

Informatization therefore seems to be closely associated with control, and the economic value attached to this control. Consequently, computers as the primary tools for informatization can be analyzed ethically in terms of the type of control that they can affect. What is deemed worthy of control by the current systems of the state and markets may differ from what the society considers as important, and can thus serve to identify spaces from where informatization should be removed, or carefully monitored, or where informatization should be increased for social good.

Weizanbaum raises other issues with unrestricted informatization. He argues against the *logical formalism* of human feelings like love and care, because such feelings if emulated by computers can reduce the agency that humans place in themselves (Weizenbaum, 1976). This is undesirable because human agency is vital especially to handle situations where computers fall short or make mistakes, and it is infeasible to informatize everything to the extent that it eliminates all uncertainty in the outputs generated by computerized systems. It is precisely such uncertainty that is meant to be handled by humans based on their values which shape their judgements. Similarly, not all social processes can be reasonably informatized. For example, the modelling of crime through heatmaps for the purpose of proactive policing has been criticized as being an improper model that not only misses out on actual factors that lead to crime but also amplifies biases which cause some areas to be more actively policed than others (Scannell, 2015). Informatization therefore may have both rationalist limits as well as moral limits in determining what should or should not be informatized. Agre's criticism of AI rests on similar grounds, that the field is not open to criticism about these limits, and to accept any criticism it demands a utilitarian justification through the demonstration of an alternate working system which is better (Agre, 1997).

To summarize, it seems that the ethical implications of computers can be understood in terms of what is informatized, how well it is informatized, and what this informatization is meant to control - labour processes, individual behaviour, social behaviour, human agency, etc. Informatization makes control easier, and control can lead to exploitation. Exploitation can further be assigned an economic value based on the surplus extracted from labour, which through the economic rationality of capitalist systems is then able to govern informatization itself. Although I do not pursue this line of thinking further in this book, it could prove useful to build a general theory of computer ethics.

4.2 An Ethical Consistency Test for Technologies

I examine three systems: Facebook, which is the most popular social networking website with billions of users; Aadhaar, which is a unique identity platform put in place by the Indian government to serve as an identification and authentication platform for many public and private services; and Mobile Vaani (MV), which is a federated network of voice-based participatory media platforms meant for less-literate users to share and discuss common concerns and solutions with one another. Aadhaar and MV explicitly position themselves as technology for social good projects, while Facebook also comes close with its normative mission to connect people.

Through these examples, I demonstrate how one can use the proposed ethical framework to conduct a values-based analysis of these platforms that uncovers the underlying values driving their design and management. I do this through the several analysis methods, such as examining the stated policy positions and mission statements, interviews of key personnel, published studies, algorithmic auditing, and news and stories in the media about the platforms. This reveals the underlying ethical values in the design elements and management practices of the platforms. These can be checked for internal consistency. Manifestation of the ethical values in terms of their salience, intention, and enactment, can also reveal the ethics-based foundation of the projects (Martin et al., 2017).

4.2.1 Mobile Vaani

MV uses interactive voice response (IVR) systems to enable even less-literate communities to share information without requiring the Internet or smartphones (Moitra et al., 2016). Community members call a phone number and the IVR servers cut the call and call the person back, making the service free of cost to the users. Users can listen to a sequence of voice messages or record their own message. Recorded messages undergo a manual moderation process, and if they meet the editorial guidelines then the messages are published on the IVR for other users to listen.

MV enables several use-cases such as question-answering, where people record questions and others (or a designated panel of experts) can answer these questions in audio form, local news reporting by citizen journalists, policy-related discussion and feedback, grievance redressal where people ask for assistance especially with getting access to government welfare entitlements, and cultural expression. Several other platforms operate in similar ways, such as CGNet Swara which is also aimed at grievance redressal and citizen journalism (Mudliar et al., 2013), Avaaj Otalo focuses on agriculture-based question-answering (Patel et al., 2010), Sangeet Swara for singing and other cultural expressions (Vashistha et al., 2015), and Sangoshthi for health workers to learn with one another (Yadav et al., 2017).

MV is structured as a federated network of multiple platforms, each having its own unique phone number and content-mix, with customized editorial policies specific to the context of the platform. These platforms are location-specific, roughly one per district in the geographies where MV operates, or theme-specific such as dedicated platforms for people with physical disabilities, for adolescents and youth, for industrial sector blue-collar workers, and for pregnant mothers and those with small children in rural areas of North India, among others (Seth et al., 2020).

MV platforms are started with specific purposive goals and a ToC towards meeting these goals (Gram Vaani, 2019):

- The choice of using IVR makes the platforms more inclusive so that non-Internet or less-literate or low-income users can participate.
- The moderation process ensures veracity of content to avoid misinformation or hate speech.
- Diversity and pluralism are supported through the editorial process by encouraging debate and discussion as long as the arguments are put forth in a respectable tone, and by featuring diverse arguments equitably so that differing viewpoints can all be discussed.
- A network of community volunteers and civil society organizations respond to grievances about welfare schemes and needs of the participating communities so that they can be provided with assistance, thus going beyond a purely informational role that media platforms otherwise typically provide.

The MV manifesto therefore highlights values such as *inclusion* to reduce inequities in access to information platforms, *taking responsibility* to ensure information credibility, *pluralism* with debates and discussion as a means to build mutual respect and understanding between diverse stakeholders, and to *support and help vulnerable* populations. These values are an integral part of the design, purposively built by intention, and enacted even with only limited access to resources (Seth, 2020d). Further, several of these values are terminal values, such as *equity* for vulnerable populations by providing them with additional support, and ensuring that diverse communities build a strong mutual understanding through pluralistic communication.

When MV was rolled out, field interactions brought out the importance of capacity building of users to avoid appropriation of the platform by more skilled users (Moitra et al., 2018). This led the MV team to place more emphasis on volunteers to conduct in-person demonstration and capacity-building workshops. Over time, this offline activity was found to introduce new biases based on caste and gender affiliations of the volunteers, which caused them to focus more on only some groups. The MV management team, therefore, started placing more emphasis on creating diversity among the volunteer-base, which in turn translated into a diversity of the user-base as well. This management practice of being in touch with the users through regular surveys, field visits, and listening to grievances of users about the platform itself demonstrated further evidence of enacting the values of inclusion and responsibility.

Similarly, editorial policies for moderation also underwent changes as new issues were noticed, such as incidents of cyber-bullying on the platform, the need for anonymity when raising serious allegations of local corruption, or to not publish certain kinds of news for the safety of the users (Seth et al., 2020). The MV moderation process itself was audited through anonymized logs of content listenership to evaluate whether the editorial process was giving near-equal exposure to different aspects of a topic under discussion. An algorithm has been developed to improve this further, by ensuring short-term diversity and long-term fairness in the ranked list of audio messages that are played to callers (Muskaan et al., 2019). Efforts are also underway to decentralize this algorithm by training community volunteers to let them locally specify algorithmic parameters to control the editorial policies of their own platforms.

4.2.2 Facebook

Unlike MV, Facebook did not start with a clear end-goal or ToC. With a vaguely stated mission to *build communities*, it also has had to deal with several allegations about the means chosen to pursue this mission. This includes issues with its working model, such as the spread of misinformation that it was unable to prevent during the 2016 US Election, and security loopholes that led to leakage of private user data in the Cambridge Analytica scandal (Cadwalladr, 2018; Mocanu et al., 2015). We focus here on the former and use the ethical framework to reason what could be the objectives, design decisions, deployment oversight, and the ethics-based foundation of the social media platform that led to such misuse.

During the 2016 US Election there seems to have been failure either in the timely spotting of platform misuse as part of the management processes, or in prioritizing the observations to quickly iterate on the design so that new features or processes could be rapidly built. News articles point to lapses on potentially both fronts (Thompson & Vogelstein, 2018). This indicates that values such as the *responsibility* of the technology provider to ensure a safe space for its users were only peripherally salient. Subsequently, in the 2020 US Election, Facebook took these issues very seriously and put better processes in place, showing a shift in salience of the values of *responsibility and safety*, and leading to the actual enactment of these values (Roose, 2021).

Facebook also does not demonstrate strong terminal values in its work. This is evident by analysing its news feed, where Facebook does not make any attempt at ensuring that users receive a complete perspective on an issue (Bakshy et al., 2015; Muskaan et al., 2019; Tufekci, 2016). It rather claims its neutrality through the argument that its algorithms do no worse than what a random selection of news from the users' social network neighbourhood would have done. Based on this argument, Facebook does not attempt to counter any echo chambers that emerge naturally because of the homophily effect in social networks (Cinelli, et al., 2021). This puts a low salience on values like *diversity and pluralism*, and the absence of any terminal values in Facebook.

4.2.3 Aadhaar

The Aadhaar system aims to provide a unique digital identity to each Indian resident, validated through their biometrics. An authentication platform is also provided to government departments and services for granting access to welfare benefits. This is believed to reduce leakages in welfare delivery by preventing the same person from being in possession of multiple IDs that could be used to get excess benefits (Planning Commission, 2019). Biometrics are also believed to be an inclusive means for authentication, not requiring people to own phones or carry smartcards. Finally, Aadhaar also advocates for linkage of identities with bank accounts for seamless payments and benefit transfers to eliminate rent seeking by government officials in providing welfare payments to beneficiaries. This hints towards values of *inclusion* and *equality* in the user interface to help beneficiaries jump the digital divide and illiteracy barriers, and terminal values like *reducing corruption* and *improving efficiency* in welfare payments.

Criticism has however been raised about Aadhaar for an incorrect diagnosis of problems. Being denied welfare for not having an identity is less of a problem than lacking documents to enrol for welfare schemes, or not knowing the enrolment procedures, and rent seeking by government officials to register citizens (Gupta et al., 2021; Seth et al., 2020). Even leakages due to identity fraud are minuscule as compared to leakages due to quantity fraud in the distribution of subsidized food grains (J. Drèze & Khera, 2015). Aadhaar cannot plug such leakages. This raises concerns about the true purpose of Aadhaar. It has been alleged by civil society activists that Aadhaar was always meant as an identification system for national security and to suppress dissent, and meant for commercial players to more easily join disparate datasets and open up rural markets for credit-related financial products (Reetika Khera, 2019; U. Ramanathan, 2017).

Other stated values have also been questioned. Extensive reports have surfaced of how Aadhaar technology design is not suitable for challenging rural conditions, which are marked by poor Internet connectivity and biometric matching errors of false negatives for older people and labourers. This has led to exclusion of millions of people from benefits, but despite strong feedback from many researchers and civil society organizations, no design changes have been made as of 2022 (R. Khera, 2017; Seth et al., 2020). This indicates an inconsistency in values of *inclusion* and *equality* that were used to justify the technology choice, but have not been considered in the management of the socio-technical interface. Hardly any pilots were done by the government before commissioning this nation-wide project, which raises further doubts on its values of *responsibility* towards the citizens. In fact, no research reports have been released by the government on the efficacy of biometrics to successfully eliminate duplicate identities, raising doubts on the most significant benefit claimed by Aadhaar (Mathews, 2016).

A similar gap is noticed with the self-service design of the system. Despite strong socio-technical evidence of the need for beneficiaries to easily obtain assistance with government procedures, the self-service design of Aadhaar makes it harder for volunteers to provide assistance. Deployment experiences show that less educated and the poor need to get help from social workers and officials to rectify mistakes in the Aadhaar data, but this is done informally, which, in fact, leads to security lapses (Gram Vaani, 2017). No changes in the design have, however, been incorporated to allow assisted usage through trusted intermediaries. This again puts into question the values of the Aadhaar designers towards *inclusivity*, *mentoring*, and *facilitating* usage for those people unable to deal with the technology, as opposed to expecting them to improve their skills or overcome these barriers.

The centralized and non-transparent machine-driven decision-making architecture of Aadhaar, with no easy appeal procedures for mistaken decision-making, has ended up putting power back in the hands of the service providers who use Aadhaar as an authentication platform. This has therefore taken power away from the hands of the beneficiaries, and is seen extensively in the use of Aadhaar for authenticating subsidized food purchases. Shop owners are able to leverage technology failure as a means to exercise power in different ways, for example, to deny supplies to legitimate beneficiaries and sell it instead in the black market, or to grant supplies as a special favour rather than as a rightful entitlement (R. Khera, 2017). This further increases the power differential between the shop owner and consumers, and can manifest itself in other spheres of community life where local elite exploit the less powerful. In the same way, subservience to a centralized decision-making system run by the government disempowers beneficiaries because of their dependency on a system that controls access to their life-critical entitlements. Such a system design clearly does not encourage values of *power-based equality*.

4.3 Summary

I have discussed so far some essential criteria that can be used to define projects for social good, using an ethics-based terminology of terminal and instrumental values that the projects espouse. These values themselves should be decided through participatory democratic means by society. Technology projects, constituted of various design elements and management practices, can be checked for internal consistency to ensure that the ethics-based foundations of all the constituent elements are consistent. Methods such as using newspaper reports, ethnographic studies, algorithmic auditing, and interviews of the project stakeholders can be used to discover the underlying ethics of the constituent design and management practices.

I now build upon this framework. In Chapter 5, I further discuss the necessity of ethics in the management practices of technology projects, being spaces where gaps often arise because ethics embedded in the project design may not be sufficient by themselves. In Chapter 6, I then further discuss a core value that especially technology for meta-social good projects should include, to bring about power-based equality in society. These chapters are specifically meant to draw attention of technologists to non-technological aspects of their projects, which are essential to ensure that their technologies do not produce disempowering effects.

Chapter 5

The Limits of Design

In this chapter, I deepen the argument that both design elements and management practices of a technology project need to be aligned with its ethics-based foundations. There are many reasons for this, as described earlier: Technology providers may be unable to control what their technology gets used for and by whom, they may not understand the limitations of their own technology in advance and the wide diversity of contexts in which their technologies may be used, or the regulatory response of the state might also be slow in drawing attention to critical ignored aspects (Angwin et al., 2016; Arun, 2019; Buchanan & Seshagiri, 2016; O'Neil, 2016; Solon Barocas et al., 2019; Vosoughi et al., 2018). Even projects that use ICTs for development and are closer to social good with technologies designed specifically to address certain social development objectives may fail in achieving their desired outcomes for similar reasons (Barboni et al., 2018; R. Khera, 2017).

The dominant approaches which have emerged to ensure responsible outcomes from ICTs are by embedding ethics into the design of the technology artefact, and to use participatory methods to design the technology so that power dynamics and other aspects about the users' context can be taken into account (Berdichevsky & Neuenschwander, 1999; Bjerknes & Bratteteig, 1995; FATML, 2019; Friedman et al., 2013; Goldsmith & Burton, 2017; Mumford & Weir, 1979; Spinuzzi, 2005). Post-colonial considerations, critical design, and subaltern approaches attempt to gain a better understanding of the social contexts, but in practice these techniques have been applied only for design or to interpret the outcomes of technology (Irani et al., 2010). I demonstrate in this chapter using the example of Mobile Vaani (MV), that ensuring responsible outcomes and social good by design alone is not sufficient. I argue that much of the harmful effects of technology arise at its socio-technical interface, when it is deployed and used by people, post-design. Careful design can enable or constrain certain affordances in how the designed objects are used, and the design can also be altered iteratively to modify these affordances (Volkoss & Strong, 2017; Vredenburg et al., 2002). However, the notion of design that a blueprint shapes the usage of the designed objects is inadequate (Blomberg et al., 1993; Ideo, 2008; E. B. Sanders & Stappers, 2008; L. Sanders, 2008; Spinuzzi, 2005).

Technology and (Dis)Empowerment: A Call to Technologists, 83–102

doi:10.1108/978-1-80382-393-520221005

I argue that the *management* of the designed objects is also important to shape their usage, especially to ensure that responsible outcomes or social good arise through their use. Although the design literature often does not distinguish between design and management practices, considering management methods as a part of design as well (Escobar, 2018), it is important to draw this distinction because in practice, the two streams are operated differently and the challenges faced by them are different. While challenges for design are more related to accurate modelling of the context so that the envisioned social good goals can be met through pre-determined pathways, challenges for management are more emergent and stem for various aspects of the socio-technical interface as I discuss in this chapter. Moreover, a lack of attention placed on management of the socio-technical interface can be misleading – approaches of ethics by design may solve some problems but not all, and relying on them only could, in fact, give a false sense of safety.

Managing the socio-technical interface can include many aspects, which I illustrate through MV. Who is included or excluded from access to technologies, policies to shape appropriate usage norms, and priority placed on social impact as compared to other issues like financial sustainability, all need careful management of the technologies beyond what can be ensured by the design itself (Barboni et al., 2018; Chandrasekharan, Samory et al., 2018; R. Khera, 2017; Lampe & Resnik, 2004; Moitra et al., 2016; Seth, 2020d). These aspects may be unforeseen at the design stage, which may sometimes be due to a lack of adequately diverse prototyping or simply due to surprises that are bound to arise when technologies are deployed in an immensely complex world. Attention to management becomes even more relevant today when many digital platforms have deeply permeated our lives: These platforms are being used by billions of people, and are embedded in a complex global web of finance and politics, so that it is daunting to even conceive of them being significantly re-designed or replaced. In such a situation, it becomes imperative to examine the management processes of these platforms to understand how problems arising at their socio-technical interface are being managed. Are these processes participatory, what ethical values underpin these processes, and are they overly bureaucratic or are they context-sensitive, are some questions to ask in this regard (Geiger, 2015).

My proposal in this chapter is that just like several design frameworks suggest ethics as the foundation to design, similarly ethics can serve as guardrails to build management processes for the management of the socio-technical interface as well. I outline several aspects of this interface that need careful management, and suggest some corresponding management practices using the case study of MV which I describe here in more detail. My arguments resonate with those by Dahlbom and Mathiassen who state that technologists need to immerse themselves as activists to manage technology operations, only then can they truly have a chance of influencing the outcomes (Bo Dahlbom & Lars Mathiassen, 2016).

5.1 What is Design?

5.1.1 Shortcomings of Ethics by Design Approaches

The need to ground technology design in ethics has been recognized since a long time. Duquenoy and Thimbleby point out the power imbalance that exists between designers and users, and that designers should recognize their responsibility to ensure that their innovations create a just world and do good (Duquenoy & Thimbleby, 1999). This becomes challenging because designers may have biases and may not always know the circumstances of their users or how the technology might affect them. Hence they suggest that designers should operate using the Rawlsian principle of the *veil of ignorance* by placing themselves as unspecified users, and then check that the technology will not erode their own rights of liberty and equality (M. Sandel, 2009). Sen's criticism of the Rawlsian framework is based on the recognition that equality will be hard to achieve as an eventual goal with just using the veil of ignorance, since disadvantaged individuals may need access to additional resources to help them catch-up with others, pointing to the need for equity-based mechanisms like affirmative action (A. Sen, 2009; Simons, 2019). Applying this logic to Duquenoy's suggestion supports our argument that design alone may not be sufficient to avoid outcomes like inequality: Users less capable to use the technology, for example, may lose out because benefits from the technology will accrue to those who are able to use it effectively. Therefore deployment management beyond the technology design, such as capacity building of less tech-savvy users, will be needed for equity and to avoid an increase in inequality (T. Unwin, 2018a, 2018b).

Things are even harder in practice. Far from following Rawlsian or other ethical frameworks, the politics of designers shape what values they espouse. Winner showed through many examples that technology design codifies the politics of designers, which could potentially manifest into societal-scale changes when wielded by powerful agencies such as authoritarian governments or corporations and the media (Winner, 1980). Processes to notice such problems, through regular feedback or other mechanisms, then become essential to take corrective action. Escobar similarly describes how the modernist and rationalist Western concept of development leads to the imposition of a flawed top-down design due to their simplistic modelling of local contexts (Escobar, 1995). Such a design needs careful management through continuous feedback. Only then can such projects have any chance of success in meeting their stated goals or to achieve responsible outcomes.

To overcome these challenges of designer bias, Participatory Design (PD) approaches were developed (Bjerknes & Bratteteig, 1995; Mumford & Weir 1979; Spinuzzi, 2005). PD methods go beyond methods like co-design and human centred design where typically external designers engage with prospective users to understand them, and build, prototype, and fine-tune tools for them (Ideo, 2008; L. Sanders, 2008). PD on the other hand is grounded in democratic values aimed at enabling all users to influence the design by arriving at a consensus through joint consultations, or discuss conflicts in agonistic spaces (Bodker, 2018). It especially focusses on power differentials between different classes of users such as when designing tools that are to be used at a factory workplace where employers and workers may have conflicting concerns. It also embraces ongoing capacity building of the users to enable them to participate in the design process effectively, and to design tools such that they lead to fulfilment of democratic objectives such as equality and justice, rather than narrow objectives to create increased usage of the technology tools (Bjerknes & Bratteteig, 1995; Christiansen, 2014; Farooq et al., 2005). It has been applied widely within organizational settings but the focus has primarily been on getting the design right, as opposed to ongoing attention being required for deployment management as well (Hirschheim & Klein, 1989; Mumford & Weir, 1979). Several long-term PD projects do discuss challenges with sustainability but have not analyzed the relevance of processes to manage various aspects of the socio-technical interface, especially in the context of digital platforms for information sharing (Lodato & DiSalvo, 2018).

A greater challenge with the full-fledged PD process is the practicality of applying it in today's context of developing and scaling large digital platforms. PD requires time-consuming consultations, prototyping, capacity building of the participants, etc., whereas platform designers are typically driven by a build-and-break approach with a singular goal to gain quick user traction so that they can lay claim to more funding to scale their platforms (Seth, 2020d). Any fundamental problems in the design that might surface later become concretized and hard to change as the platform grows larger, making it difficult to control technologies when they get more popular (Collingridge, 1980). Further, the commercialization strategy of platforms may keep them from following democratic principles deemed essential by PD and cause them to align with certain types of users. This is clearly seen with Uber which favours commuters more than drivers, or blue-collar work platforms which seem more interested to serve employer interests than worker interests (Betterplace, 2020; Rosenblat & Stark, 2016).

The value sensitive design (VSD) approach has a similar starting point as PD in terms of following a participatory process to arrive at a core set of values through consultations with users or other means (Friedman et al., 2013). These values are then baked into the technology design so that they are never violated when the technologies are deployed. Based on several VSD examples discussed in the literature, however, this typically seems to have led to the incorporation of context-free design features like data privacy or fairness definitions or informed consent to be encoded in the technology (Winkler & Spiekermann, 2018). When technologies are deployed in diverse contexts, however, and demand dynamic adaptation to new situations, then alertness in management processes becomes essential to change the design, or to manage the affordances allowed by the design. Such discussions are absent in the VSD literature. I discuss in this case study of MV many examples that needed careful management irrespective of the initial design methods.

Note that I am not proposing a design method as an alternative to PD or VSD or even other methods. My intention is to simply point out that most design methods have not looked closely at deployment management. In fact, design methods such as PD and VSD would indeed be a first step to technology design, with deployment management as the subsequent step, followed by ongoing interactions between the two. As clarified in the previous chapter, a common ethical system can serve to guide both the design and deployment management of technology.

5.1.2 Action Research for Deployment Management

How deployment management can be done to steer technology? The action research method comes closest (Greenwood & Levin, 2007; G. Hayes, 2011). Action research has more ambitious goals than PD and aims to continually shape an intervention based on deployment feedback, with all decision-making done through participation of the community in the process.

This naturally requires long-term engagement with the community, with slow and careful evolution through experimentation and consultation. Such an iterative process is however perceived as unwieldy and impractical, and truly long-term action research-based interventions are rare even in the ICTs for development space (Best & R. Kumar, 2008; Bodker & Kyng, 2018; Braa et al., 2004; Cross et al., 2019; Koradia et al., 2012; Surana et al., 2008). This returns the emphasis to greater responsibility lying with the managers of technologies to ensure responsible outcomes when technologies are deployed. In forthcoming sections, I use the MV case study to outline several useful processes that can facilitate careful management of the socio-technical interface during deployment, and come close to action research methods.

5.1.3 Need for Deployment Management for Artificial Intelligence (AI)

Concerns arising with the application of AI in many domains have of late brought significant attention to the development of new data management and algorithmic techniques. Biased and discriminatory decision-making arising from unchecked biases in the data used to train machine learning algorithms, has led to the adoption of several methods to audit the data as well as organizational processes to ensure that such auditing is compulsorily done (Angwin et al., 2016; Bird et al., 2019; O'Neil, 2016). Research communities have realized that the objectives codified in algorithms can also lead to biases, such as the choice of fairness definitions to be used in different applications, or ranking algorithms that can amplify biases, and regulators have also begun to pay attention to algorithmic auditing and the need for explanatory capability in algorithms (Solon Barocas et al., 2019; Chouldechova & Roth, 2018; Diakopoulos, 2017; Goodman & Flaxman, 2017; Hindman et al., 2003). Participatory processes to collect inputs from users to accordingly choose and parameterize AI algorithms, have also seen progress (M. Lee et al., 2018).

These developments are rooted in an ethics by design model. To preserve continued ethical functioning of technologies will require due attention paid to deployment management as well. For example, deployment management processes are needed to ensure that any biases in the data are detected and steps are taken on priority to re-train models so that they perform in line with the committed accuracy and fairness guarantees (Bird et al., 2019). Similarly, accepting accountability for the outcome of the algorithms, transparency and explainability of the results, and providing appeal procedures against unfair decisions made by the algorithms, are necessary to deal with mistakes and take corrective action. Downstream issues arising from different contexts where AI algorithms may be deployed, similarly require a continued engagement (Solon Barocas et al., 2019; K. Vaccaro & Karahalios, 2018). All these demand careful management of AI systems by humans, rather than to eliminate humans for time or cost efficiency. As suggested by Floridi, AI is not intelligent by itself and humans need to retain their agency rather than submit to AI (Floridi, 2018). Ensuring ethics by design in AI technologies, as adopted in declarations such as ICDPCC, are therefore unlikely to be sufficient by themselves (ICDPPC, 2018).

5.2 Case Study: Mobile Vaani

I next describe the case study of a voice-based community media platform called Mobile Vaani (MV) which has been operating since more than a decade in rural central India (Moitra et al., 2016). I highlight several values that shaped the design of the platform technology, and I then show how these values also guided management of the socio-technical interface of the platform, including substantial changes in the scope of the project as it added new programme components beyond the technology to meet the project goals. I discuss several aspects of the socio-technical interface that threw up surprises as MV was deployed and scaled, and the processes developed to manage these aspects. I do not consider MV as a pure PD or action research project because most decisions were not taken in consultation with the users, but were guided through a focussed understanding of the context and iterations on the design or deployment management policies accordingly. The MV case study therefore shares more in common with digital platforms like Facebook and Twitter, where, as discussed in previous sections, the designers and managers of these platforms have greater onus to ensure responsible outcomes from their technologies, since participatory or action research-oriented methods are not practically viable for them to follow. Similarly, MV has been used in diverse contexts by different classes of users who could sometimes be in conflict with one another, and has been subject to financial and scaling pressures that tend to put social impact objectives on the back-burner. I therefore believe that many of the processes developed to manage the MV socio-technical interface may be generic enough to be applied to other digital platforms as well.

5.2.1 Background About MV

The concept of community media is centred in the idea that the needs of local or interest-based communities are best understood by the community itself, and editorially driven mass media run by external institutions is not able to address these needs largely because of their lack of local context. Community media thus enables user autonomy for communities to define their own agenda and create and distribute their own content. Several such models for community embedded media initiatives exist in India, such as Khabar Lahariya which supports rural women to write newspapers, Video Volunteers, which trains rural community correspondents to produce videos about human rights, and several community radio stations that create radio programmes of local relevance (Khabar Lahariya, 2020; Pavarala, 2007; Video Volunteers, 2020). In this rich mix of initiatives, MV started as a phone-based community media service in rural central India in 2012. Driven by a value to be inclusive towards less-literate and low-income populations, MV uses interactive voice response (IVR) systems to enable voice-driven communication and allows access even via simple phones that cannot access the Internet. Several other initiatives have also reported similar successes with using IVR systems in rural environments (Koradia, 2012, 2013; Mudliar et al., 2013; Patel et al., 2010; Vashistha et al., 2015; Yadav et al., 2017).

MV works as follows: People give a call to a well-known phone number and hang-up. The IVR system then initiates a call back to the callers, effectively making it a free service for them. The people then use phone keypad buttons to browse a list of audio messages or record their own message. Any messages recorded by them go through a manual moderation process before being published on the IVR for others to hear, essentially to filter out poor quality audio recordings or inappropriate content.

The goal of this initiative was to bring about social development through the use of technology, along the following lines.

- First, taking a human rights-based approach which values dignity and equality of power, the MV team was convinced that such a platform could improve transparency and accountability in local governance. MV would make it simpler for people to share information about civic or governance problems that they were facing, and to demand transparency in decision-making.
- Second, prior research in social media analysis showed that discussions on participatory media platforms are useful because homophily leads to people receiving information from their strong ties which is highly contextual and understandable by them, and participation by diverse users from across weak ties leads to information that enhances completeness (Granovetter, 1973; McPherson et al., 2001; Seth et al., 2015). Messages evolve as they traverse a social network, gaining both context and completeness through comments by people occupying different positions in the social network graph (Seth & Zhang, 2008). It was, therefore, believed that since MV could facilitate people to share their views and experiences on various topics, it could lead to a more complete and actionable understanding of relevant topics, and thereby also support the values of plurality and diversity for deliberation and learning.
- Third, people gain significant self-empowerment when they can be heard by many others, that is, they gain a voice, and are able to challenge exploitative power structures, thus strengthening the values of dignity and equality (Pavarala, 2007).

- Fourth, strong community bonding happens when community media hosts traditional songs and music, and covers local events, making the community closer knitted (Koradia et al., 2013).
- Finally, while mobilizing these above pathways, the MV team also wanted to have a business and operations model that could be replicated readily for scaling either directly or through partners.

MV was, therefore, initiated with a clear design of what terminal and instrumental values were to be embedded in the technology and project, and the change it was supposed to bring through the pathways listed above. The actual journey turned out to be more complex! By building processes to manage the socio-technical interface, guided by the same values incorporated in the technology design, the MV team has been reasonably successful in adhering to their founding vision.

5.2.1 MV's Socio-technical Interface

Recall that the socio-technical interface is the boundary between technology, engineering, user interface, and system design, on the one side; and factors that shape how people use this technology, on the other. I do not mean to suggest that people who engineer the technology do not consider the social context. In fact, several of their design choices may be shaped by the social context very much like how the MV choice of using IVR was shaped by the low literacy and high mobile phone usage in the communities of interest. The socio-technical interface, however, that I am interested in here is in the dimension of how the technology is used by people *post-design*. The different aspects of this socio-technical interface which I discuss next are by no means exhaustive, but nevertheless a reasonable starting point. The rest of this section is written chronologically to serve as a historical record.

Aspect #1: Technology Literacy and Access. Soon after starting, the MV team realized that IVR was an alien concept for many people (Moitra et al., 2016). In general, though many men had access to phones and were able to use it to dial numbers and pick up calls, most had not used automated IVR systems in the past. The usage of the system had to be demonstrated to them, and in the absence of any other media in these areas, offline training sessions were found to be the most effective (Koradia et al., 2012). Further, they had to be explained the concept of community media and what they could do with access to such a platform. That too was not straightforward to convey – it required practical examples and self-validation for people to understand the use of the technology (Moitra et al., 2016). Finally, access to phones by women and access to women to tell them about MV, were both significant challenges. Due to patriarchal norms, women are typically less literate, do not own their own phones, rely on shared access, and consequently their capability to use phones is lower than that of men (Barboni et al., 2018; D. Chakraborty et al., 2019). Access by women to public spaces is also lower and hence it is harder to reach women to inform them about MV. It was clear that the wide usage of mobile phones was not going to easily translate into MV adoption, and low-cost and scalable processes needed to be developed to overcome these barriers.

Related work has explored several interesting dynamics through which new technologies are learned. Poorly literate construction workers were able to learn a complex sequence of steps to share videos over Bluetooth, suggesting that self-motivation to use technology (in this case, for entertainment) could lead to self-learning (Smyth et al., 2010). In a study of Facebook use among urban youth in India, a mix of financial and social incentives led to users teaching their peers about the platform (N. Kumar, 2014). In the context of women, despite strong patriarchal norms, some women in urban areas learned to navigate family and community spaces to use mobile phones (N. Kumar, 2015). Similar dynamics were seen with MV as well. Users would tell their friends about it especially if their message got published on MV. Some would listen to MV in groups which led to wider listening and learning. Hearing stories of validation of MV's impact, especially in the area of grievance redressal, which I describe later, also led to MV achieving strong social credibility and popularity through word of mouth (Moitra et al., 2018, 2019). These were, however, not systematic and reproducible processes that could be considered as part of a replicable MV model. The MV team, therefore, took a different approach, as explained in the next section, to develop an innovative low-cost offline process via community volunteers which could manage the need to create technology and platform literacy.

Aspect #2: Community Embeddedness. We felt that MV's agenda should be driven by the community itself, including for content creation, the choice of discussion topics, etc. Related work has discussed the concept of *communitization* of technology, that is, when a community learns the essence of what a technology can do and is able to leverage it for the community's needs (Marsden et al., 2008). A strategy recommended to achieve communitization is through identifying a few key tech-savvy community members. Termed Human Access Points (HAPs), they are essentially technologically advanced users who understand both the needs of their community as well as the capabilities of the technology. They are able to conceptualize relevant use-cases for the technology, and also mentor other community members.

As MV was introduced into new communities, we kept running into HAPs who had quickly understood the technology and were among its early adopters. For MV to gain popularity, people were inducted as community volunteers from among these HAPs, and a financial incentive model was built to cover out-of-pocket expenses incurred by them to popularize MV and guide its usage. They would travel to different villages and tell people about MV, demonstrate its usage, and encourage its adoption. The volunteers were also encouraged to discover use-cases for the platform on their own, based on their understanding of the community (Moitra et al., 2018). As a result, over time, when MV expanded to different locations running their own local MV chapters, the volunteers created their own topic priorities for respective local MV chapters. The volunteers in one location, where MV was popular among farmers, built linkages with local agricultural institutes to answer agriculture-related questions put up by farmers. Another location built linkages with school and college coaching classes to advice their predominantly youth user base with career-counselling tips. All locations also took up a hyper-local news reporting function which was soon discovered to be a universal need, due to the scarce penetration of other media in these geographies. Such processes helped MV adhere to values of user autonomy to run community media platforms that were governed by the people.

The platforms were supplemented with common content created by the central MV team on cultural and entertainment themes, discussions on social norms such as early marriage and domestic violence, rules and eligibility for government schemes, etc. The choice of this content was guided by feedback processes such as through user interviews over the phone, focus group discussions, and also IVR-based surveys to get regular feedback (Moitra et al., 2016).

The journey with identifying and training volunteers, and retaining them, was however by no means smooth. I next discuss the aspect of internal accountability expected of the volunteers to build local MV chapters into sustainable institutions in themselves.

Aspect #3: Internal Accountability for Sustained Participation. Initially MV started as a state-wide service in the state of Jharkhand and was popularized in different regions within the state by the volunteers. Subsequently, services were also started in the states of Bihar and Madhya Pradesh. The MV field team, however, consisted of only a few personnel and who found it difficult to coordinate with dozens of volunteers from across different locations. Further, volunteer attrition became high because many people would join MV as volunteers with the expectation of financial returns, but the small stipend that was offered was not attractive (Moitra et al., 2018). The MV team realized that they had to improve their selection process to identify volunteers who were genuinely interested in bringing a positive change in their communities and for whom social incentives would be stronger than monetary incentives. The concept of federated groups in the context of trade unions is believed to be more resilient than having a single large group, and the same method was adopted for MV (Olson, 1965).

A federated MV network was established by splitting the state MVs into district-level chapters, each of which had their own unique phone number and could build their own identity. Further, the volunteers from each district were grouped into a volunteer club for that district, as shown in Fig. 5.1. Each club elected a coordinator and all the club volunteers met on a monthly basis to discuss their activities and plans. Through this hierarchical arrangement, the MV field team now only had to engage with the club coordinators. It was also found that this structure built strong solidarity and mutual accountability among the volunteers, which reduced volunteer attrition rates to practically zero. A careful financial incentive model was developed as a combination of a group incentive which was divided equally among all the volunteers in the club (calculated pro-rata on the number of active users in each club) and individual incentives for each volunteer (based on the number of good quality contributions by the volunteer, and offline community mobilization activities organized individually by them) (Moitra et al., 2018). This model further made explicit the ethos that a volunteer club should act as a collective to which all the volunteers were expected to make contributions to achieve the club's collective aim. It also helped the MV team realize the importance of collectivism as a value, which since then has influenced several other decisions as well. The club model further helped construct the kernel for

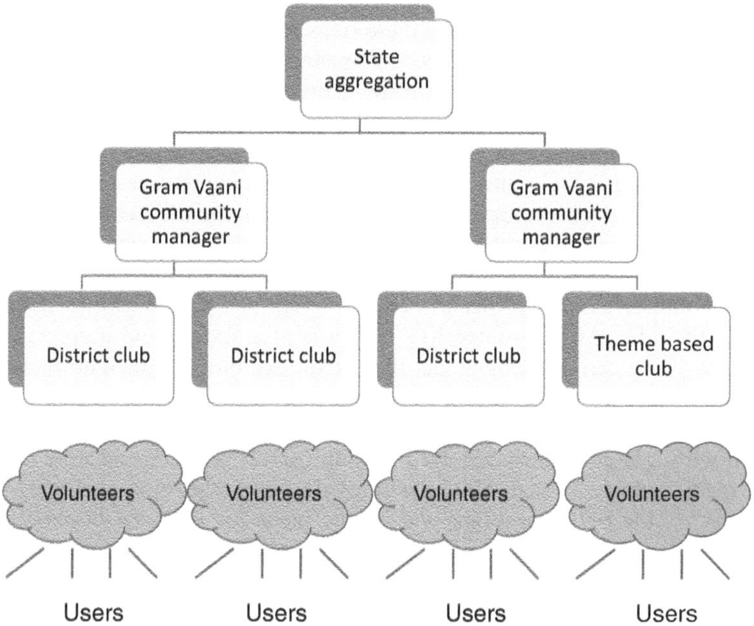

Fig. 5.1. Federated Network of MV Platforms.

MV operations that could be readily replicated for scaling, and has already been tested with replication at 30 local chapters.

Aspect #4: Biases in Inclusion and Exclusion. The unique position of power occupied by the MV volunteers raised some unanticipated issues as well. The volunteers would sometimes popularize a version of MV's mandate that made more sense to them based on their individual socio-economic–political views and priorities, which could be different from that of their clubs. They would, similarly, sometimes prioritize access for a select group of users by training them well while excluding others. For example, during the initial days of MV, several volunteers were associated with human rights activist organizations and hence were more interested in governance topics, to the extent that they began to discourage people from using MV for cultural expression through folk songs and poetry (Moitra et al., 2018). Similarly, they would sometimes discourage users from recording content themselves and would record it on their behalf, especially when these users were from less-educated backgrounds. Inherent social norms also caused disruptions – once a class-based conflict arose in a club when a lower-class volunteer was elected as a club coordinator, thereby challenging historical norms of power.

Such incidents have now become rare due to MV's more rigorous selection and training methods, and also its paying special attention to recruiting volunteers from diverse class and caste backgrounds. The underlying value of plurality helped guide the MV team in taking these steps. What was also useful in making the necessary course corrections was an openness to hear complaints that the users recorded on MV or shared with the team during field visits. This internal grievance redressal process helped empower the users, and helped the MV team to uncover new and emergent cases of undesirable appropriation of technology, again highlighting the relevance of continuously listening to user feedback.

A challenge that still remains unsolved is in overcoming the technology gender divide (Barboni et al., 2018). Most of the MV volunteers are men and find it hard to reach women to tell them about MV. Having a female volunteer in a club of all-male volunteers is also difficult in the dominant patriarchal cultural setup of rural north India. An all-women MV club was also started and despite all the volunteers being extremely dynamic, the active user base of the club remained small due to the limited mobility of women volunteers to reach other women (Moitra et al., 2018). It is worth mentioning that in a recent project the MV team worked with a large women Self Help Group (SHG) network and were able to reach many women through SHG meetings (D. Chakraborty et al., 2019). The regularity of the SHG meetings, which take place for financial bookkeeping, and the exclusive women constituency to which access was gained, provided an opportunity for both targeted outreach to women as well as repeated interactions with them encouraged them for technology adoption. Gaps still remain, though. For instance, meetings with SHGs of very poor women are held irregularly since these women were busy with work or often migrated to other locations, and hence those who could potentially benefit the most from access to the platform could not be easily included.

These examples highlight the importance of managing the socio-technical interface to prevent biases in inclusion and exclusion that can arise due to the social context or due to appropriation of the technology by more powerful and adept users. Not tackling this challenge stands the risk of entrenching existing inequalities because more powerful or tech-savvy users will be able to leverage the technology for their own agenda and move further ahead, leaving the others to perpetually catch-up (T. Unwin, 2018a, 2018b).

Aspect #5: Nurturing Responsible Usage Norms. Like any other social media platform, MV is also susceptible to misuse through submissions of fake news or hate speech content (Arun, 2019; Mondal et al., 2017). User-generated content on MV is therefore manually moderated (Moitra et al., 2016). Moderators can also download and edit the content to improve its audio quality, add transcripts and tags to the content, and control its ranking on the IVR. Content moderation serves an important role of signalling to users about what is permitted or not, and thereby shapes usage norms, as has also been seen in studies on Reddit and Slashdot (Chandrasekharan, Pavalanathan et al., 2018; Chandrasekharan, Samory et al., 2018; Lampe & Resnik, 2004). As of now, an average of only 0.5% of rejected contributions on MV are due to objectionable content, showing that users rarely attempt to misuse it. The bulk of rejections happen due to unpreparedness in recording well-articulated content, for which offline training or manual phone calls are made to guide users.

Whenever misuse has occurred, it has been dealt with severely. Cases of hate speech are escalated to volunteers who directly speak to the people. Abusive or threatening recordings can even be reported to the police. In general though, hate speech or angry voices against other users have been rare, and even discussions on contentious topics have taken place in a measured tone of respect and decency that honour values of plurality, dignity, and mutual respect. A liberal moderation policy allows as many voices as possible, and filters to reject content are based chiefly on concerns about audio quality and the tone of the message. This seems to indicate that, in contrast to many Internet-based social media platforms such as Facebook and Twitter, which allow unmoderated postings and then resort to algorithmic policing or community reporting methods, misuse is prevented on MV by establishing a precedent of respectful use from the very beginning. Volunteers and users are passionate about preserving this space and vociferously protest if objectionable content sometimes slips by the moderators.

Editorial policies to ensure diversity in content are also actively pursued. The moderators rank content based on its quality and novelty, by prioritizing content which is more detailed and informative, or brings a new viewpoint (Moitra et al., 2016). Careful curation encourages diversity. A content selection and ranking algorithm has also been developed to automatically guarantee properties such as short-term diversity and long-term fairness in representing various aspects of a given topic (Muskaan et al., 2019). Crowd-sourced indicators may also help to ascertain the value of different pieces of content, as suggested in several experiments on community radio and other voice-based forums (Koradia et al., 2013; Vashistha et al., 2015).

Aspect #6: Building Social and Institutional Credibility. Sustained participation and use of a technology will only happen once users trust it and their expectations with the technology are met. I next discuss the relevance of managing user expectations so that the technology can gain the respect of users; another important aspect of the socio-technical interface.

To understand the specific community needs around which MV framed community expectations, I first give an overview of the context in which MV is deployed. Regional newspaper and television media have not reached most MV communities, and also have a chequered reputation of ignoring problems of lower caste people, or suppressing news against the local elite possibly in return for extortion payments (Ninan, 2009). Poor and marginalized groups have, therefore, historically lacked a strong voice in the community, and even their political representation has had its own ups and downs (Mosse, 2018). They are also intimidated by complicated bureaucratic procedures for grievance redressal and may not be able to take time off their daily routine to pursue even legitimate cases of violation of their rights and entitlements (D. Chakraborty et al., 2017).

In such a context, MV is presented to the community as helping meet several needs. Media related use-cases are conveyed through statements such as: ‘It is a platform for you to talk about whatever you feel is relevant for you and your community that is not covered in the mainstream media’, ‘You can get breaking news about your community well before any newspaper or TV channel reports it’, and ‘MV is a platform where you will get useful information about agriculture, career counselling, health, government schemes, among other topics'. Governance related use-cases were predominantly conveyed as: 'You can discuss local and national policy, and we will convey your feedback to the right stakeholders', and 'MV volunteers will help resolve problems that your community is facing, especially on welfare entitlements and public services'.

While these use-cases emerged in a bottom-up manner through the involvement of volunteers, positioning MV with these strong promises stood the risk that if expectations were not met then people would dismiss MV cynically as yet another false promise. MV was, however, able to gain significant social credibility by successfully demonstrating its impact, which helped people validate its stated promises and intentions, and also demonstrated the value of honesty. The editorial processes of moderation were kept liberal. Towards the initial stages of MV, the moderators even made phone calls to users who seemed to be wanting to say something important but were not able to articulate it well, and guided them to make better audio recordings. This helped validate a strong commitment of MV towards empowering users. Similarly, grievances recorded by the people, or questions asked by them, were rigorously followed-up with reminders and support provided by the moderators, to keep users informed about actions having been initiated upon their request. All MV volunteers were also trained to cover any news events in unbiased ways, and strengthened people's perceptions about MV as being an unbiased news source about local events.

MV was also able to successfully facilitate improvements in local governance. A detailed analysis of social accountability loops formed by MV was done based on 300+ impact stories (D. Chakraborty, 2018; D. Chakraborty et al., 2017). This validated that the target beneficiaries of welfare schemes need offline support via social workers to access entitlements because self-service mechanisms like centralized helplines and web portals are hard to use. Furthermore many categories of grievances arise due to local issues and cannot be resolved centrally. MV's network of volunteers not only provided offline support to people in engaging with government authorities, they also used MV to draw the attention of government officials to resolve certain strategic grievances, which led to higher success rates of grievance redressal.

An IVR-based survey of over 500 MV users across the states of Jharkhand, Bihar, and Madhya Pradesh showed that 67% of the users agreed that MV is different from other mass media in giving an opportunity for anybody to voice themselves, 69% acknowledged the value of dialogue created on the platform to understand different viewpoints, 88% reported an increase in connecting back to their cultural roots, 85% reported an increase in political awareness, 50% acknowledged having learned new ways to articulate their views, 64% reported having gained agency in addressing problems with local governance directly themselves, and 84% acknowledged strong offline support received from MV volunteers in helping solve their problems (Moitra et al., 2019).

These functions of grievance follow-up and news reporting, in addition to exposing users to technology literacy about MV, demonstrate the relevance of deployment-management processes beyond the design stage, and also show how these processes emerged from similar ethical values which had shaped the design.

Institutional credibility was also important for MV to get constructive reactions from the state for grievance redressal or on policy implementation feedback collected from the users. As MV volunteers built stronger networks with local government officials, and demonstrated sustained usage over many years, the state too responded positively and began to view MV as an innovative means for citizen engagement which they could utilize themselves, while respecting the independence of MV. Local government officials now routinely use MV to make announcements of new schemes and subsidies, give interviews about their views on policy implementation, provide a commitment for resolution of community issues, and respond actively to requests by MV users and volunteers. Institutional credibility thus reinforced social credibility, and helped embed MV not just within the community but also with other local institutions. This is clearly a direct outcome of MV going beyond just functioning as a technology platform, to also nurturing its relationships with multiple stakeholders while upholding its underlying values.

Aspect #7: Influence of the Business Model. The final aspect I discuss is complexities that arise from the MV business model. MV has three potential revenue streams (Seth, 2020d). First, philanthropic donations and government advertising to fund awareness and behaviour change campaigns on topics such as health, nutrition, education, and livelihood. Second, community funding where the users may themselves contribute small amounts to enable and access a community media platform, plus crowd-funding to raise microgrants for sponsorship of specific activities by economically well-off people. Third, commercial advertising by companies interested in reaching rural markets. While there is strong validation of the first revenue stream, the second is untested as of now, and the third is yet to be tested at scale. I believe that all three revenue streams will emerge sooner or later, since MV presents a good product-market fit in the absence of other media outreach channels for Indian rural markets. What is not known, however, is what challenges they will present to the unbiased and community-driven coverage currently provided on MV. As revenue streams from government or corporate advertising get larger and MV's dependence on them increases to ensure its financial sustainability, MV may become susceptible to have its agenda influenced by government and corporate interests. The MV team does not have any experience so far in building processes to manage this likely forthcoming challenge, since MV has until now been sustained either through philanthropic grants or internal funding by Gram Vaani, but it is likely to emerge in the future as an important aspect of MV's socio-technical interface and I believe it will need similar grounding in values.

5.3 Managing the Socio-technical Interface

I showed in the previous section that several complex aspects of the MV socio-technical interface had to be managed to ensure responsible outcomes, beyond the initial design of the technology itself. Even the original theory of change required many changes to ensure that the project met its stated goals while still adhering to the underlying ethical values of the project. This active management of the aspects during deployment highlights a responsible handling of the project by the MV team. Similar aspects of the socio-technical interface are known to exist on other digital platforms like Facebook and Twitter, for example, when groups of users appropriate the platforms to their advantage, the platform fails to sustain its values over time, there is a failure in the emergence of good usage norms, there is a reduction in the perceived credibility of the platforms, etc. We showed that it is possible to build processes to manage the socio-technical interface and ensure responsible outcomes by taking guidance from underlying ethical values. In their absence, technologies which were meant to empower people may actually disempower them. I next attempt to generalize these processes.

5.3.1 Management Through Processes

Federated Platforms. By establishing a federated network of district-wise MV clubs that operated autonomously, but with a uniform oversight applied on their operations by the MV team, MV was able to effectively service locally relevant use-cases and create a strong ownership among its volunteer base to sustain these services. Similar methods can make management of digital information platforms easier, and allow for contextualization and diversity in the use of technologies. Smaller communities can evolve use-cases according to their needs which can lead to greater community embeddedness of the technology. Further, this community embeddedness can be facilitated through a subset of users who can mediate as volunteers or community representatives, with whom technology providers can engage in more detail. This can help make centrally managed platforms more participative (Geiger, 2015).

Internal Feedback Processes. The MV team paid close attention to issues raised by users on the quality of the service or when it was misused, and also regularly conducted user surveys and field studies to understand user needs more deeply. Technology providers of digital platforms can similarly build processes to get regular usage feedback on factors such as the following.

Internal grievances: Being able to listen to grievances raised by users can alert technology providers to misappropriation of the technology by malicious users, or malfunctioning of the technology itself, both of which are likely to happen in any digital platform. Addressing these issues promptly can contribute to building platform credibility, and also create loyalty among the users to not have the platform get misused.

Tracking of inclusion and exclusion: Biases in access or usage arising from gender or other categories of inequity can be spotted through periodic demographic studies of the users. The reasons behind the biases can be identified through deeper field immersion and critical theory inspired approaches for analysis, and appropriate action can be taken. Inequities in the social context that may lead to such inclusion and exclusion biases can thus be avoided.

Design of Incentives. Managing federated setups and building a closer feedback loop with the end-users may require additional effort from technology providers. MV showed that it is possible to distribute this effort by building appropriate incentives to involve the users or their community representatives. Suitable social, solidarity, and monetary initiatives can lead to strong internal accountability and sustained usage, and create long-standing community embeddedness. Further, being able to align the incentives with positive social change as an overarching objective may lead to greater participation of users in the management of digital platforms.

Addressing Gaps in Technology Literacy. Not all users can be expected to have a good understanding of how to use any technology and what can be achieved through it. Tracking inclusion–exclusion biases in user demographics and understanding interactions patterns to distinguish between power users and less tech-savvy users, can help spot biases that could arise from this disparity in technology literacy. Appropriate steps can then be taken to bridge the gaps. Depending upon the context, this gap could be bridged through online mechanisms like tutorial videos or it may require offline mechanisms such as training workshops, and it may be undertaken by the technology providers or it may require incentive mechanisms for the users or community representatives. In some extreme contexts, bringing about this technology literacy may not be straightforward or even possible, and in such cases processes to facilitate assisted usage may be needed. All such steps can significantly avoid the disempowerment effects that some sections of users may notice from their inability to learn to use new technologies.

Signalling. Misuse of technologies can be avoided by nurturing appropriate usage norms. Methods like content moderation or highlighting positive behaviour are useful to send signals to the users about acceptable and unacceptable practices.

5.3.2 Examples of Other Platforms

To summarize, I have discussed that unforeseen aspects at the socio-technical interface can surface even when well-designed technologies are deployed. These aspects can include the need to address gaps in technology literacy and access, increase community embeddedness, achieve internal accountability, guard against technology appropriation, shape appropriate usage norms, achieve strong social and institutional credibility by meeting expectations of the users, and create a business model that can bring financial sustainability without compromise. Technology providers need to incorporate management processes to manage such aspects of the socio-technical interface for their technologies, and underlying ethical values can serve as guardrails to guide this activity.

Consider a mobile-money company that wants to reduce fraud on its platform by malicious users who take advantage of unsuspecting and less technology-savvy customers. Stories of such phishing incidences are common (Edmund, 2015). To manage this aspect, some project teams in the mobile-money company may want to run financial literacy awareness workshops with their customers, while other teams may want to find technological solutions to detect fraud, and yet others may suggest doing nothing and letting the customers learn on their own. A common underlying ethical system can help suggest which of these approaches should be chosen - prioritizing equality and inclusion may prompt the first response, an ethics by design approach would lead to the second pathway, and a belief in markets and individualization may lead to the third option. In the case of MV, an ethical value of inclusion was followed consistently by the MV team that led to steps to build offline volunteer networks that could reach marginalized user groups. Not prioritizing this value could have led to an altogether different solution, like focussing on low-hanging fruit to on-board only young male users who are already technology-savvy and can be acquired at lower costs (Moitra et al., 2018; Sachitanand, 2018).

Although organizations have always had teams to manage their technologies, my emphasis in this chapter is on uncovering the ethical values that drive the development of these management processes. This is not a well-studied concept. Barring the discussion of content moderation as a deployment management activity (Chandrasekharan, Pavalanathan et al., 2018; Chandrasekharan, Samory et al., 2018; Lampe & Resnik, 2004), most literature has focussed only on characterizations of misuse, such as along the dimensions of gender (Barboni et al., 2018), technology access (R. Khera, 2017), and information veracity (Arun, 2019; Vosoughi et al., 2018). The study of deployment management of technology as an area, it seems, has not had as much attention as the study of design. Most design innovations are analyzed only at a prototype stage. The study of prototypes, however, misses out on complexities at the socio-technical interface that emerge in long-term deployments.

I next describe a few other digital platforms and the fallouts when strong processes are not built to manage their socio-technical interface. One of the largest digital platforms, Facebook, has lost considerable credibility in recent years for the limited attention it paid to check the presence of echo chambers and filter bubbles created through its algorithms, the metrics it chose to optimize, slow efforts to detect fake news, and poor ability to control data leaks (Arun, 2019; Cadwalladr, 2018; Tufekci, 2016). Platforms such as Reddit and Slashdot on the other hand, have shown resilience to such cases of misappropriation (Stoddard, 2017). They have a strong moderation system mediated by people, and Reddit in fact is set up as a federated system which allows different communities to build their own respective moderation policies (Chandrasekharan, Samory et al., 2018). A number of tools are provided to make moderation easier based on upvotes and downvotes given by the users on stories and comments, by helping identify controversial entries or biased entries. Such self-regulation by individual forums has given Reddit strong credibility that its forums do not suppress information but rather engage, discuss, and debate, to identify good quality content that should be featured prominently (R. Mills, 2011). Stringent moderation practices such as banning of users who attempt to misuse the platform, and algorithmic improvements to bring more fairness in news ranking, send strong signals to users to avoid misuse (Chandrasekharan, Pavalanathan et al., 2018). Slashdot, similarly, has a hierarchy of content moderators who ensure accuracy in the news and views accepted for publication, and further mechanisms exist to mentor new moderators (Lampe & Resnik, 2004). Other successful collaborative knowledge building platforms like Wikipedia and Quora also have strong moderation policies and a somewhat implicit federated structure defined on the basis of topic interests of users.

These examples show that managing the socio-technical interface of participatory media platforms is indeed possible by allowing humans to make editorial decisions and using technology as a tool to aid them in this process, rather than entirely delegating decision-making to technology. It has similarly been suggested that strong institutions created by the users themselves can arrest misuse on platforms like Facebook (Freuler, 2018). Even though Facebook has now scaled its human-driven moderation processes, along with automated methods to detect misuse, it allows only very coarse moderation features and thereby does not truly empower users to take responsibility for the administration of their groups. While users can flag objectionable posts, the subsequent process of how these posts are handled, and reasons why they were flagged in the first place, are not made transparently available which dampens the signalling mechanism to build more responsible usage practices (Crawford & Gillespie, 2016). In contrast, platforms like Wikipedia, where debates on acceptance and rejection of edits are publicly available, are more agonistic spaces that also serve as editorial learning grounds for new users and to signal appropriate usage norms (Crawford, 2016; Halfaker et al., 2014). Like Facebook, Whatsapp also does not provide adequate tools to the administrators to manage their groups and prevent misuse. Further, the encrypted nature of communication makes it difficult for other stakeholders, including Whatsapp itself, to manage it. Whatsapp has, in fact, absolved itself of any responsibility towards facilitating more appropriate administration of its forums. I therefore conclude that even though such platforms have enabled communication and collaboration at massive scales with significant benefits for society, not managing their socio-technical interface effectively has led to undesirable consequences.

In contrast, going back to Reddit, we see an example of having strong user-led institutions that can influence platform management policies. Volunteer moderators of many popular Reddit groups demonstrated a deep personal investment in the platform during the 2015 AMAgeddon episode when Reddit fired an employee who was very popular among the moderators (Centivany & Glushko, 2015). The moderators marked their dissent by making many subreddits go dark and demanded active participation in Reddit's administration. Such user involvement in technology management is closely related to the notion of appropriate technology introduced earlier in Chapter 2. Schumacher highlighted that a vital aspect of appropriate technology is for it to be controllable by local communities (Schumacher, 1973). While he was primarily referring to locally manufactured mechanical devices, such as agricultural implements, the insight is generalizable even to information and communication technologies. Being able to steer and manage technologies helps users to understand their capabilities and limitations, and to strengthen local institutions that can design management processes to control the technologies. Although securing participation and consensus in governance is not straightforward, studies of appropriate technology and commons-based management principles offer important insights (Heeks, 1999; Ostrom, 1990; Pedersen, 2015; Vines et al., 2013). This is further discussed in Chapter 8.

5.4 Summary

I have tried to show in this chapter that management of the socio-technical interface of technology projects in terms of underlying ethical values is important to ensure responsible outcomes and whether the technology project is meeting its stated social good goals or not. The framework proposed in Chapter 4 to evaluate the ethics-based foundations of the project objectives, its design elements, and management practices, can be used to check the consistency of the ethical values at different layers against the social good goals of the project. While these are generic principles that can be applied to information technologies to evaluate their ethical foundations, in the next chapter I return to the question of which values should social good, or meta-social good projects, adopt. In Chapters 2 and 3, I briefly introduced the salience of power-based equality as a terminal value to counter many social relationships of domination and exploitation that technology can bring about. In the next chapter, I delve deeper into this concept and argue that the value of power-based equality should be considered as a core value of technology for social good projects, and further shape the values chosen as part of the project objectives, design, and management elements.

Chapter 6

Ensuring Power-based Equality

Chapter 3 discussed why doing social good is a purposive concept, and that an ethics-based terminology can express social good because it provides useful principles and reasoning. Given the purposive- and hence consequentialist-direction of social good, an ethics-based framing must have at least some terminal values or constitutive freedoms specified as end-goals. Chapter 4 described a framework for technology projects as being constituted from several design and management elements, each having their own ethical foundations, and that cross-checking the underlying ethics of these components against the social good definition of a project can be used to spot conflicts.

I now expand these arguments further by asserting that achieving *power-based equality* needs to be a core terminal value to define social good, to build a society in which power is shared equally. This is based on the observation that technology systems are both shaped by and can shape power relations between different stakeholders (Foucault, 1984; Winner, 1980). Power may emerge, for example, through an accumulation of capital which can be used to influence policies as in neoliberalism, or production relations in industrial units where managers can control labour processes, or through bureaucratic systems of governance that impose authoritative structures on people and their behaviour, or even control over knowledge and ideology that is shaped by institutions such as schools and the media. These power relationships of domination are anti-humanist and create social alienation. They are usually further accentuated through technology, which tends to reflect and magnify the virtues and vices of social systems that created it. Neutralizing these power relationships to bring power-based equality is therefore essential, and this end-goal is especially relevant for meta-social good projects, which aim for a transformation of the current systems of governance. Since meta-social good projects are then meant to define the values essential for social good, power-based equality also becomes indirectly important as a core value in other social good projects.

I build on the extensive literature on the theorization of power to propose how power relations can be modelled in both the design and management of

Technology and (Dis)Empowerment: A Call to Technologists, 103–129

Published under exclusive license by Emerald Publishing Limited
doi:10.1108/978-1-80382-393-520221006

technology systems, and evaluated against whether these elements help increasing power-based equality. This narrows what can be claimed to be social good, and introduces a political element, since power relationships are ultimately social constructs, and technology systems that aim to influence power relationships thereby take an explicit political stand. This is not a shortcoming! In fact, it is essential in the current global context, as described in Chapter 2, so that a clear distinction be drawn between what is social good and what is not, and accordingly people – consumers, technologists, and citizens, alike – can make informed choices. The modelling framework I propose below for power relationships is also useful for technology projects to consider how power relationships between direct and indirect users can be shaped, and avoid any further disempowerment of weak stakeholders (Avelino et al., 2017; Seth, 2020c). This modelling framework can also be considered as a means for technologists to incorporate critical theory about power relationships and exploitation into the design and management of their technologies.

I first present a brief survey of the concept of power. I then argue that power differentials of various forms are at the root of many problems that social good projects should address, and therefore power is a useful lens through which to view social good projects. Technology can play a dual role in either removing power differentials in society, or entrenching them. Therefore, I argue that a technology project can legitimately said to be doing social good only if it is examined through the lens of power and it aims to reduce power-based equality among project stakeholders through its design and management components. I then propose a modelling framework to understand the network of power relationships in which a technology project may operate and which it aims to alter. This framework can help project designers and managers to analyze their projects, take decisions, and especially spot instances where projects claiming to do social good may actually be disempowering. Finally, I demonstrate the modelling framework through several examples.

Power also implies responsibility and consequently being accountable for the outcomes. An additional benefit, therefore, of examining technology projects through the lens of power is to know where to impose accountability when harmful outcomes arise. Accountability can lead to liability, which can aid responsibility (Nissenbaum, 1996).

6.1 What is Power?

Although the concept of power has been extensively debated (see Lukes, 2004 for an extensive survey), it has found a consistent practical use in social development, where entrenched power differences between individuals and groups impact the success of social development projects. This provides an appropriate grounding for social good projects that have similar goals. Power has been conceptualized as being of different *types*, and operating at different *levels*, discussed next.

6.1.1 Types of Power

A common negative type of power is *power-over*. This denotes individual actors or groups or institutions such as the state, which, through formal rules or informal societal norms, hold *power-over* others in curtailing their freedoms (Pettit, 2013). Authoritarian states or despotic regimes or gender and caste norms may exercise power through active coercion or behaviourial control. *Power-over* is therefore a relational concept between a pair of actors where one dominates over the other.

Positive types of power include *power-to, power-with,* and *power-within. Power-to* is similar to agency, and denotes the power that an individual has to do something, such as to exercise a freedom without having to seek somebody else's permission or approval (Batliwala, 2019). *Power-to* is a relational concept between an actor and a task or freedom. *Power-with* is about collective power, to confront injustice through mobilizing and joining hands with others facing similar injustices. *Power-with* is therefore a relational concept between multiple individuals deriving their power from one another, and also like a resource which emerges and gives the individuals the *power-to* accomplish something they otherwise may not have been able to accomplish alone. Both the *power-to* and *power-with* concepts are typically applied by individuals and groups to negate the power that somebody else may have *over* them. In contrast, *power-within* describes the sense of confidence, dignity, and self-esteem that comes from a person gaining awareness about their situation and realizing the possibility to do something about it. *Power-within* is thus a resource that is a precondition to take steps to exercise the *power-to* take action, or to identify *power-with* others (Pettit, 2013).

6.1.2 Levels of Power

The different types of power may operate at different levels (Hardy & Leiba-O'Sullivan, 1998). At the *first level* is power that operates through the possession of certain resources, like wealth and knowledge. An actor A may have *power-over* another actor B because A is more knowledgeable than B, and therefore B may be dependent upon A to accomplish some tasks. A could use the power they derive through this differential to coerce B towards doing something, or deny the ability for task completion to B by refusing to share their resources. If B is able to acquire similar resources then B would be able to exercise their own *power-to* accomplish the tasks, or B could join hands *with* others to acquire collective power that would bring the necessary resources. At this level, therefore, the absence of the *power-to* do something, or absence of the opportunity to gain *power-with* others, can lead to those possessing the necessary resources to gain *power-over* others.

At the *second level* is power that operates through decision-making rules. Norms and laws may put rules in place that enable some individuals or groups to impose *power-over* others. Gender norms such as patriarchy or caste norms that discriminate or constrain access to various spaces, are examples of such norms where a group is able to exercise *power-over* other groups. Having the *power-to* change these rules and norms, or acquire this power by joining hands *with* others, can counter the power that dominant groups may hold *over* other groups. These rules and norms also shape resource acquisition at the *first level* at which power operates, such as constraining females or lower-caste groups to challenge power holders by restricting their educational opportunities to acquire knowledge, or their mobility to be able to join others for collective action to acquire resources. Rules for bureaucratic procedures may, similarly, give veto authority to officials, which then act as a resource at the *first level* for them to hold *power-over* other citizens.

The *third level* is about how rules are legitimized, that is, the ideology behind the rules. Democratic setups aim to give equal *power-to* everybody to shape the ideology, but cultural hegemony, or the management of meaning through the media, are methods through which dominant groups continue to hold *power-over* others (Foucault, 1984; Gramsci, 1971; Herman & Chomsky, 1988). The media shapes what people think about (agenda setting in media) and also how they think about it (framing bias in media) (Scheufele & Tewksbury, 2007). Power can thus shape knowledge, and thereby influence the sense of ethics and morality that people perceive, to want to alter these power relationships (Foucault, 1984). This has been highlighted by Martha Nussbaum through several examples, such as many women subjected to domestic violence lacked any sense of being wronged, or women doing hard labour in brick kilns saw it as normal to not wanting to learn new skills while men did so to get promoted (Nussbaum, 2000).

A *fourth level* is also discussed but is contested (Lukes, 2004). Foucault mentions that power is everywhere, and even those in power in some contexts may be powerless in others, resulting in local effects where entrenched power relationships are sometimes disrupted. These differences in local context that lead to power being challenged in certain spaces can, however, be modelled using the first three levels and do not require a fourth level to be defined. This is similar to James Scott's concept of anarchy, where weaker groups may concede and comply with power in visible venues but counter it through hidden and untraceable means in less-visible contexts (Scott, 2012). Although such hidden means to alter power relationships may not be easily visible, they can be modelled among the first three levels. For our purposes of understanding power, I therefore only consider the first three levels at which power operates, through the four types of *power-over*, *power-to*, *power-with*, and *power-within*.

To summarize, power at various levels may, therefore, arise through different sources, such as resources (material wealth, knowledge, official position, and self-confidence) at the *first level*, norms and laws at the *second level*, and management of these rules at the *third level* by institutions such as the Parliament, media, judiciary, and the executive.

Some additional typologies to understand power are part of the Powercube framework (Gaventa, 2019). The operations of power may be *visible*, *invisible*, or *hidden*. For example, ideology shaping power at the *third level* is typically hidden, such as corporate influence on policy-making (A. Sen et al., 2018). At the *second level*, formal stated laws may be *visible*, but social norms may be *invisible* because they are rarely spoken about (Spivak, 1988). The institutional setup to change rules may also be characterized as decision-making spaces that are *closed, invited,* or *claimed. Closed spaces* are opaque where decisions are made behind closed doors. Some decision-making spaces could *invite* people to participate as citizens, beneficiaries, or users. *Claimed spaces* are where collective action is practiced to bring rule making into the open.

6.1.3 Overcoming Power Differentials

While most analysis of power studies how *power-over* emerges and is observed, I argue that social good should be concerned with how these power differentials can be overcome (Lukes, 2004). For example, development programmes such as sports for girls, mentorship for income generation, and financing through women Self Help Groups, aim to foster self-esteem and self-confidence to build greater *power-within* (Pettit, 2013). Welfare schemes, such as for universal health coverage, the right to education, and the right to employment, aim to provide basic resources to everybody through which they can possess the *power-to* utilize other opportunities and resist the power that those with these resources may hold *over* them (A. Sen, 2000). These schemes operate through rules, and the rules can be shaped by democratic mechanisms such as elections to incorporate affirmative action so that historically marginalized groups are supported more equitably. Collective action (*power-with*) is the primary means to alter such rules.

Collective power in the space of labour rights distinguishes between *associational power*, *structural power*, *institutional power*, and *societal power* (Schmalz et al., 2018). *Associational power* is acquired when workers at a workplace associate with one another to form a union, or multiple unions come together in an industry sector or across several industry sectors to represent workers. When this *associational power* is put to use, it is called *structural power* at the workplace or in the labour market. *Workplace bargaining power* exercised through strikes and sit-ins is able to influence management to provide better working conditions in a factory or company. The legitimacy itself to collectivize and protest through worker unions is secured through the labour laws in a country. *Marketplace bargaining power* is similarly able to influence policy within the existing labour laws to increase minimum wages, provide stronger social security, and even influence labour supply under certain types of worker association arrangements (Jatav & Jajoria, 2020). Marketplace power may also be exercised through industry-wide or country-wide strikes and sit-ins, or through political affiliations of unions to mobilize their demands by strategically siding with different political parties.

Both *associational* and *structural power* can be considered as operating at the *first level*, as forms of the *power-with* type of power. When *associational* and *structural power* is stronger, it may lead to *institutional power* to change the rules themselves, at the *second level*. These may take the form of introducing rules for co-determination in companies that gives workers more say in the management of the company, or changing the rules to strike or associate in a workplace, or rules to mandate new forms of social security. *Societal power* emerges at the *third level* when broader society, beyond just the workers and unions, is able to represent the interest of the workers. This typically takes the form of coalitions between different social movements such as labour, human rights, and feminism, etc., becoming allies due to their shared understanding of a common ethic and morality.

The success and failure to alter collective power relationships can be understood through political economy (Acosta & Pettit, 2014; DFID, 2009). This can help understand who participates in networked movements, their motivations, why certain rules were put into place and who they benefit, what ensures cooperation between the rule-makers, etc. Free rider effects as unions get large, and a lack of observability to identify rule-breakers or an inability to impose sanctions against the rule-breakers, are able to explain some failures in collective action (Olson, 1965; Ostrom, 1990).

Central to realizing different forms of collective power is also communication (Dijk, 1989). Communication may take the form of discourse through worker newspapers that strengthen the *associational power* of workers, and mobilize them to exercise their *structural power*. Communication may also serve to identify strategic pathways to gain *institutional power*. Communication and discourse are also essential to draw legitimacy for a common ideology that social movements can endorse to build *societal power*. Altering the ideology requires the existing ideologies to be spelled out in detail, and show how the "group cognitions influence the social construction of reality, social practices, and hence, the (trans)formation of societal structures" (Dijk, 1989). A discourse analysis of advertisements, laws, propaganda, etc., is suggested as a method to understand current ideologies, and then create new discourse to displace these ideologies. Being able to communicate itself requires resources, such as logistical resources to create and distribute media, and knowledge of regulations to overcome censorship, or the ability to use surreptitious communication channels instead (Hirsch, 2016).

The discussion so far highlights that power operates through relationships of access to resources, rules that shape these relationships, and which in turn are shaped by ideologies. Central to overcoming negative power relationships of domination at different levels is collective action as a means to alter these power relationships, and the importance of communication to this process. We show next how this relates to doing social good, and that the outcomes of projects are governed by the network configuration of power relationships within which the projects operate, and which they attempt to shape purposively or inadvertently.

6.2 Power and Social Good

Many undesirable outcomes that may arise from technology, have, at their roots, a manifestation of negative power relationships: dominant actors imposing *power-over* others through the technology; an inability of people to exercise positive power relationships because they do not have the *power-to* use technology, or they do not have sufficient *power-within* themselves to act upon the opportunities opened up by the technology; or the technology is not suitable to build *power-with* others through solidarity and collective action. We discuss this next, arguing that achieving power-based equality in access to resources (*first level*), application of rules and the resource distribution outcomes arising from these rules (*second level*), or of ideological influence to shape the rules (*third level*), should be included among core terminal values of social good projects. Social good projects should be examined through the lens of power to discover how they shape power relationships, and whether they purposively attempt to impose power-based equality of various forms in these relationships.

We demonstrate this argument by analysing power relationships in three domains: achieving the SDGs, building civic technologies for citizen–government engagement, and countering or altering capitalist ideologies to evolve greater cooperation among people (Escobar, 2018; Scholz, 2016; SDGs, 2015; Wrobel et al., 2019).

The *power-to* relationship for actors to realize their goals is closely related to Sen's concept of freedoms for development (A. Sen, 2000). SDGs such as no poverty (Goal 1), zero hunger (Goal 2), good health and well-being (Goal 3), having quality education (Goal 4), clean water and sanitation (Goal 6), and decent work (Goal 8), are essential capabilities for people to have the freedom to realize their goals. Looking at these freedoms from the lens of power shows that dispossession of the *power-to* accomplish certain goals makes people vulnerable to be dominated by those who possess this power, leading to the emergence of negative *power-over* relationships. For example, when phones or online mechanisms are the only means to access government benefits, people who do not have the basic literacy to operate the Internet or phones or even the means to own phones, become beholden to volunteers, agents, or rent seeking by government officials. Similarly, people who do not have good health to take up certain types of work end up with restricted work options. To preserve job security they may concede power to their employers and not speak out against poor wages or working conditions. In the same way, people who do not have easy access to clean water would need to spend time and energy to secure water, which can impact their ability to realize capabilities such as education or health, and which in turn can lead to power imbalances.

The lens of power thus helps reveal the interdependencies in networks of *power-to* and *power-over* relationships, as well as inequalities that can emerge if some *power-to* relationships are not equally distributed among everybody. In the examples above, if the *power-to* get quality education is not available to everybody then *power-over* relationships will still persist. Similarly, if the *power-to* having good health is not equally available then it can lead to unequal *power-over* relationships in workplaces especially those using manual labour or with a heavy reliance on productivity metrics. Inequality in the *power-to* access clean water can lead to inequalities in the *power-to* access quality education and good health, and which in turn can lead to other inequalities.

SDGs on gender equality (Goal 5) and reducing other forms of inequality (Goal 10) are also directly related to the removal of *power-over* relationships. Gender inequality fundamentally arises from social norms like patriarchy, that give men the *power-over* women in many domains such as controlling women's access to public spaces, prioritizing boys education over that of girls, and maintaining land records in the names of men (Hunnicutt, 2009). The power lens has been actively used in the gender space, and the concept of building *power-within* is seen as a route to overcome the power of men *over* women (Batliwala, 2019). Inequalities arising from race and caste have similarly been examined through the lens of power of how the more dominant races and castes are able to systematically hold on to their power by continued discrimination in access to jobs and education, and differentiated dignity of different types of work (Tilly, 1999). Collective action of the *power-with* type has been used as a means to overcome these negative *power-over* relationships. *Associational power* among lower-caste groups has often manifested into *structural power* exercised through vote banks in electoral politics, and has led to *institutional power* to put in place affirmative action rules in multiple domains (S. Vivek, 2014). Income and wealth inequality also cause the rich to have *power-over* the poor, when the power elite are able to influence policy that favours retention of the status quo through low taxation rates for the rich and erosion of social protection for the poor (C. W. Mills, 1956; J. E. Stiglitz, 2012). Collective action through labour movements and farmer protests of the *power-with* type are examples of resisting domination. The elite do not want these movements to challenge the status quo and therefore attempt to weaken the movements by using their power to shape laws or call upon the state to make it harder to conduct these social movements (Hensman, 2011). The capitalist ideology with its emphasis on individualization and competition has also weakened *associational power* among workers to build strong labour movements, or link multiple movements into a cohesive social movement for *societal power*.

These examples show how achieving the SDGs require having to navigate and alter a network of power relationships. To achieve the SDGs, appropriate *power-to* resources need to be equally distributed among everybody, or *power-with* and *power-within* relationships need to freely emerge through rules that are equally applicable to all. We can thus see that the SDGs can be examined through the lens of power to remove top-down *power-over* relationships, or build bottom-up *power-to*, *power-with*, and *power-within* relationships. The lens of power also helps reveal that SDGs interact with one another by building or breaking power relationships, and hence they should not be considered as isolated goals in themselves. Understanding the underlying power dynamics, as what critical theory aims to do, can be useful to especially develop interventionist models by formulating theories of change based on how networks of power relationships may temporally evolve as a result of these interactions.

I next look at the domain of civic technology projects which are also meant for social good. Many civic technologies aim to improve citizen–government engagement in areas such as access to welfare schemes, access to justice, civic amenities, and public safety. The trust placed by citizens in governments is reduced if they are not able to access some of these fundamental services, and can lead to growing apathy and distancing between citizens and the state which can impact democracy (Carswell et al., 2019). Overly bureaucratic procedures to access government schemes and services have in fact been termed as *structural violence* (Graeber, 2015). Red tape in filling out multiple forms, following complex procedural pathways that are not made clearly visible to the citizens, and non-responsiveness in keeping citizens updated about their queries or grievances, renders people powerless when dealing with state institutions. This absence of a *power-to* understand procedures or influence them, in turn gives rise to agents and intermediaries who facilitate engagement for a fees, or bribe-seeking government officials, who use their *power-over* the citizens to deliberately withhold completion of their applications unless the bribes are paid. Civic technology projects aim to counter these negative *power-over* relationships that bureaucrats have over the citizens, by creating new power relationships through technologies that provide the citizens with an access to grievance redressal procedures, avenues through which they can demand accountability for prompt and meaningful resolution, and performance statistics for internal monitoring of government departments. These new mechanisms thus aim to give citizens equal *power-over* government officials to counter the *power-over* relationships that the officials have over them. However, as seen in the case of Aadhaar, discussed in Chapter 4, the use of technology not appropriate to the context, or with design and management limitations, prevented these counter–power relationships from being fully realized. New types of negative power relationships are instead introduced that are regressive towards the citizens. I therefore argue that the goals of civic technology projects to improve citizen–government relations will only be realized if when examined through the lens of power, the counter–power given to citizens through these projects is made adequately functional and effective. Failing this, civic technologies may further disempower the citizens.

Many social good projects are also in the broad domain of countering the dominant capitalist ideology of individual self-interest. These include innovative concepts such as platform cooperatives for ride sharing, where drivers are owners of the platform through which ride allocations are made (Scholz, 2016). Similarly, farmer collectives that help farmers pool their lands, rotate crops, and cover for one another's damages, are more resilient to crop failures and also more environmentally sustainable than when each farmer acts in their own individual interest (B. Agarwal, 2018). Such collectives are a means for small and marginal farmers to resist the pressures of capital to dispossess them. They are able to build greater trust and solidarity, bargain effectively with upstream and downstream stakeholders in the agricultural value chain, and coordinate for sustainable and resilient use of community resources (Bosc, 2018). Another example is of independent media projects that are crowd-funded or community-funded, and try to challenge the capitalist ideology by demonstrating that alternatives to corporate controlled media are indeed feasible (Jian & Usher, 2014). Such social good projects that aim to counter the dominant capitalist ideologies are, however, often thwarted by having to compete with alternative capitalist projects that sustain the status quo. These capitalist projects are able to access funds from the organized elite, or offer lower prices to consumers by not accounting for negative externalities, and are thereby able to crowd out genuine social good projects that try to challenge the capitalist ideology.

The enduring dominance of the current capitalist ideology of private property, entrepreneurship, and meritocracy can also be examined through the lens of power (Piketty, 2020). On the one hand, the capitalist ideology argues for *power-to* forms of equality for actors to have equal access to innovation and growth opportunities through market institutions (Hayek, 1944). Inequalities however do undeniably exist in markets (K. Marx, 1867). They can be offset through *power-with* forms of cooperation among market actors, but the individualistic competition nurtured in market institutions weakens such *power-with* emergent forms. The capitalist ideology reduces the value placed upon cooperation to create fairer conditions for market operations by instead emphasizing individual meritocracy and competition (Turchin, 2018). It also diminishes the role that regulation should play in addressing market inequalities by claiming that such inequalities do not exist in the first place (Crouch, 2011). This ultimately hides the fact that power-based differentials do exist in society, and also resists regulatory or collective action forms to fix these differentials, essentially enabling the durability of the status quo. Corporate control of the mass media further serves to entrench these beliefs (Arsenault & Castells, 2008). Examining the contemporary capitalist ideology through the lens of power thus demonstrates how its ideological hegemony is able to resist changes.

Concepts such as the pluriverse also urge to recognize multiple ideologies that should be allowed to co-exist with one another, and to especially counter the dominant rationalist and modernist ideology, which, through development projects, has often produced further exploitation of the poor (Escobar, 2018). These movements too, however, struggle against the hegemony of rationalist knowledge perpetuated by the current institutions of governance, economy, and education that tend to devalue culture and social relationships, and objectify nature as a commodity separate from humans. These entrenched power relationships which do not allow other forms of knowledge to flourish, make it difficult to transform current systems. They make it harder to engage society to demand ethical consumer preferences in markets or suitable regulations in democratic states so as to facilitate the institutionalization of new values into capitalism itself (A. Sen, 2000).

To summarize, power-based analysis can thus be applied in many domains including the SDGs, civic technology, and questioning capitalist ideology. Social good projects attempt to alter existing networks of power relationships and aim to bring power-based equality of various forms at various levels. The SDGs for education and good health aim to bring an equal distribution of these resources among everybody (*first level*). The SDGs against gender and other forms of inequality will succeed only if they can foster *power-with* and *power-within* relationships among groups to counter dominant *power-over* relationships (*second level*). Civic technologies similarly need to examine whether the new rules mobilized by them enable citizens to gain equal *power-over* relationships to counter the *power-over* relationships that government officials have over them (*second level*). Social good projects such as platform cooperatives or meta-social good projects such as independent media can be seen as challenging the dominant capitalist ideology by introducing alternatives that may someday become popular enough to equally shape rules (*third level*). The need for power-based equality therefore should be acknowledged as being essential at various levels in the form of resources, rules, and ideology, based on whichever forms may be relevant for a particular project.

I therefore assert that power is an overarching concept and any social good project should include as a core terminal value the requirement to achieve power-based equality of whatever form that may be relevant to the goals of the project. Either the goal of the project itself may be to bring about a power-based equality of some form, or power-based equality will be essential to realize the goals of the project. This is not to argue for a universal set of values that should be embraced since different social good projects can endorse their own constellation of values and end-goals, but power-based equality of various forms is likely to be an overarching requirement without which other values will be unachievable (Hoven, 2010; A. Sen, 2009). This is also not to suggest that ensuring power-based equality of resources, rules, and ideology, in a particular social good project, will alone be sufficient to bring about social good. Many other factors are relevant, but considering power-based equality will be necessary to realize project goals.

6.3 Infrastructures of Power

Technologies play an important role in shaping networks of power relationships. Technologies can make some resources more easily accessible, enforce rules for equal access to resources, and provide tools for mass communication to challenge dominant ideologies. Technologies can also restrict access and tighten control for those who already hold dominant power relationships. I next describe four ways in which technologies shape power relationships: between the technology designers and users, between managers and users, between different users, and between the technology itself and its users.

Most analysis of power in technology design has studied power differences between designers and users. Designers are acknowledged as having *power-over* users due to their deeper understanding of the internal functioning of technology (Duquenoy & Thimbleby, 1999). The *veil of ignorance* originally proposed by Rawls is suggested as a means for designers to consider themselves as users, so as to not design a technology which they themselves would not want to use. Other methods such as participatory design, discussed earlier, focus on reducing power differentials between designers and users by involving the users in the design process.

Second, the nature of the technology itself is important in shaping power relationships between managers, workers, and users. Plato discussed the hierarchical command structure that needs to be in place for steering a ship, which then shapes power relationships between the sailors based on their functional relationships of dependency with one another (Winner, 1980). Engels also wrote about the nature of complex technologies that need skilled workers to operate them, which affects the power relationships between workers, and leads to authoritative structures. Mumford made similar arguments in distinguishing between authoritative and democratic technics, depending upon whether the technology needs centralization for its creation and operation or not (Mumford & Weir, 1979). Technologies with centralization characteristics thus lead to authoritative power structures between managers and workers. Lessig's analysis of the Internet architecture also demonstrates how it shapes power relationships between commercial enterprises, governments, and technology companies on the one hand, and users on the other. The distributed end-to-end architecture of the Internet that supported free sharing of content, free speech, and anonymous communication, is now threatened through regulations that aim to fundamentally alter the architecture with centralizing elements that make it feasible to perfectly trace content, expression, and communication (Lessig, 1999). This tilts the balance of power significantly in favour of surveillance by the government, or by corporations who have much to gain by tracking the online activities of people.

Taking the example of Bentham's prison panopticon, Foucault called such surveillance setups as disciplinary technologies of power: technology infrastructures that use the *power-to* punish to bring about individualization and homogenization, leading to a normalization of behaviour that produces *docile bodies* (Foucault, 1984). This occurs through two processes. *Observational control* by the prison guards over the prisoners is exercised by restricting the prisoners from having the *power-to* break routine or find access to escape routes, and to not have recourse to collectivizing to gain *power-with* others. This then leads to *regularizing control* when the desires or hopes of the prisoners to ever counter the dominant *power-over* relationships of the guards are indefinitely diminished. Participation by different stakeholders in the governance process to shape the architecture, and in the management process to develop appropriate policies, can contribute towards reducing these power differentials.

Third, technology shapes power relationships between different sets of users who directly or indirectly use the technology. The Aadhaar technology discussed earlier shapes power relationships between the beneficiaries of welfare services, service providers, government officials, technology intermediaries, and the state (Seth, 2020c). The operation of the technology is not transparent and beneficiaries are not able to understand it, which enables service providers such as fair-price shop owners to dominate over beneficiaries. To fix technological errors that beneficiaries may be encountering, they need to take help from intermediaries or government officials. This opens up other routes of domination over the beneficiaries. Different power relationships would have prevailed in the absence of the technology, but new ones were introduced through it (R. Khera, 2017). These experiences of the beneficiaries with the technology and other stakeholders ultimately shapes the relationship they perceive with the state, such as whether it is empathetic to their circumstances, or whether it empowers or disempowers them, and thereby regularizes the citizens' perceptions of control by the state *over* them, to eventually impact the quality of democracy that emerges (Carswell et al., 2019; Seth, 2020b).

Similarly, the nature of reputation algorithms on social media platforms such as Twitter shape power relationships between different users by effectively giving some users a gate-keeping *power-over* others. This happens when algorithms estimate the influence or social capital of users based on engagement metrics on content shared by them, which results in positive feedback loops and catapults some users to celebrity status, similar to *Googlearchy* dynamics caused by the page-rank algorithm discussed in Chapter 2 (Tufekci, 2016). This ultimately gives rise to gatekeepers and new forms of agenda setting on social media, which may end up drowning legitimate voices and accelerate misinformation, leading to the subversion of democratic institutions (Seth, 2019b; Treem et al., 2016).

Offline intermediated platforms, such as Mobile Vaani (MV), have been wary of such gatekeeper effects that can arise from inherent social inequities. As discussed in the previous chapter, we took appropriate steps to ensure that skilled users from higher caste and class groups were not able to appropriate the platform to promote their own agenda (Moitra et al., 2018). Giving an equal opportunity for marginalized groups to voice themselves was instrumental in shaping power relationships between different groups, through the assistance of technology.

Finally, technologies may form direct power relationships with users. For example, algorithmic decision-making in domains such as access to welfare service leads to the technology gaining *power-over* people (O'Neil, 2016). As described in Chapter 4, methods such as explainability of the algorithmic decisions, the ability to appeal against them, and the introduction of human-in-the-loop methods, can provide to people counter–*power-over* technologies. Opaque grievance redressal mechanisms may, however, leave scope for discretion by officials, and human-in-the-loop mechanisms may degenerate to mere rubber-stamping mechanisms (Jorna & Wagenaar, 2007; Wagner, 2019). Quantitative and qualitative auditing mechanisms can ensure that such mechanisms actually equalize power relationships and not just give a semblance of bringing power-based equality (Diakopoulos, 2017; Solon Barocas et al., 2019). Understanding the effects of technology on the distribution of power, and consequently to place accountability on those in power, is therefore not straightforward, and technologies indeed can have politics of power embedded within them (Winner, 1980).

However, just as technologies can entrench negative power relationships, they can also enable positive power relationships. Manuel Castells has analyzed the emergence of collective forms of *power-with* through networked communication platforms (M. Castells, 2016). He defines four different types of power. *Networking power* is about having access to resources (*first level*) to access the communication platforms and bypass gate-keeping strategies. *Network power* emerges from the existence of conducive rules, or the ability to leverage rules in desired ways (*second level*) to operate communication platforms. For example, being able to determine protocols of what is allowed or not on the network, such as by editorial policies for content moderation on participatory media platforms. *Networked power* emerges from participation on the network once appropriate protocols for network operation are in place, and is related to different forms of *associational power*, *structural power*, *institutional power*, and *societal power*. *Network-making power* is wielded by network architects (designers, programmers, and regulators) who build these communication networks and provide the constitutive elements and vocabulary for specifying protocols (*third level*). Emerging from the underlying ideologies behind the technology, this last form of power is related to the affordances offered by technology platforms for users to create and run their own groups on the platforms. The configurability provided on the platforms determines whether or not they can support certain kinds of rules and protocols (Seth et al., 2020).

These different types of powers can assist in coordination and information sharing for collective action by groups that otherwise were not able to communicate easily. With easier communication, mediated through conducive rules that are configurable on these platforms, the groups are able to acquire *associational power* which can be leveraged for *structural power*, or even *institutional* and *societal power*. Global movements such as #MeToo on Twitter for gender rights, national movements such as the Arab Spring to restore democracy in many countries, and local movements such as campaigns on MV by women against illegal liquor shops in their villages, are examples of *power-with* forms that can emerge with the assistance of networking technologies (Moitra et al., 2019).

Although the role that networking technologies have played in strengthening democracy has been widely proclaimed (M. Castells, 1996), this role has also been contested by others who argue, instead, that like any other technology, networking technologies also have ultimately played in the same power arena and have hence not succeeded in fundamentally altering its structure (F. Webster, 1995). They argue that social media platforms are used more heavily to suppress dissent than to strengthen movements against the status quo, since the affordances in their design allow their appropriation more easily by those who are already powerful (C. Fuchs, 2013). The Indian government has made dozens of arrests under its archaic sedition law for social media activity that criticizes the government (D. Agarwal, 2017). While similar views are held about the social shaping of technology based on how new technologies are appropriated by society, I believe that social good projects can break away because doing social good is purposive and directed (MacKenzie & Wajcman, 1999). Technologies for social good may be constantly forced to regress towards the status quo, but I believe that determined and careful navigation and management, with a better technology design, can, nevertheless, push these projects towards envisioned goals. Just because meeting goals is hard does not mean they should not be in place, or attempts should not be made to meet them. A similar motivation underlies critical design principles, or views by Tim Unwin that technology should be designed only for the poor, otherwise they will only amplify existing inequalities (Iivari & Kuuti, 2017; T. Unwin, 2018a, 2018b).

If understanding and designing in the context of power relationships are indeed central to doing social good, as I have argued, and power-based equality is essential to realizing any meaningful success, then the ethics-based definitions of social good need to have power-based equality as a core terminal value. This value needs to be realized throughout the various design elements and management practices of the technology. I next discuss how networks of power relationships can be modelled to describe the system and to track whether power-based equality is indeed enabled and is able to change entrenched power structures.

6.4 Modelling Power Relationships

Social good goals can only be meaningfully achieved if power-based equality is achieved through the project. Being able to model and visualize the network of power relationships in which projects operate, and which the projects aim to alter, can help project designers and managers to analyze, reflect, strategize, and course-correct the direction of their projects It can help reveal who is being empowered and who is not, through the technology projects. In this section, I describe a modelling methodology that aims to make legible the networks of power relationships and how they may change. I draw upon modelling methodologies such as Actor-Network Theory (ANT) (B. Latour, 1996), power relationships (Gaventa, 2019; Pettit, 2013), systems thinking (Meadows, 2008), and cybernetics (Wiener, 1950). I suggest a methodology to express the system rather than predicting its behaviour; behaviour prediction models can potentially be built on top of the framework and is a subject of future research. I do outline some system archetypes, however, that usually lead to undesirable outcomes, and can be spotted in the models.

6.4.1 Background: ANT and Systems Thinking

ANT studies the role that technologies play in structuring social relationships. It provides many useful formalisms. The notion of an *actant*, as opposed to an actor, emphasizes the active role that actors play when they are paired with resources (B. Latour, 1996). Many actants can also combine to form a network, which, once stable and *black-boxed*, can be considered as an actant unit in itself. This helps simplify the examination of networks. Until an actant-network is not stable, it keeps transforming or *translating*, and it is this analysis of *translations* that ANT facilities in social theory (B. Latour, 2005). Several devices are suggested to understand these transformations. Actants can try to firmly establish themselves in a network by becoming indispensable for various activities, and are then called Obligatory Points of Passage (OPP) (Callon, 1986). Technology too can be an actant, and can become an OPP if its relevance in the network is significant. The process through which OPPs entrench themselves in a network may involve the *enrolment* of allies, especially those who speak on behalf of the OPPs. The term *immutable mobiles* is used to refer to marketing scripts that actants and their allies may use during the *mobilization* process to insert themselves as OPPs in the network. ANT therefore takes into consideration individual interests and incentives of actants, how actants acquire agency through their resources, their structural position in a network, and the processes they follow to alter the network to achieve their goals. It fundamentally provides an economic lens to understand network translations.

ANT also emphasizes that social asymmetries cannot be ignored (B. Latour, 1996). These asymmetries lead to challenges at the different stages of *enrolment* and *mobilization*, and bring inherent uncertainty in how networks evolve. It is here that ANT appears promising in analysing how power relationships shape network transformations. However, ANT does not characterize different types of power relationships and how they may affect the network. There is room, therefore, for the introduction of the typology of power relationships introduced earlier, to potentially be layered on ANT. For example, this can help understand why certain types of ANT enrolment may succeed (due to *power-over* relationships between actants) or fail (due to resistance gained through *power-within* or *power-with* relationships), and lead to different distributions of *power-to* relationships.

Another shortcoming of ANT is that while it helps explain why some networks are stable and others are not, by understanding the aligned interests among the actors, it does not allow a modelling of overall network objectives. It is essentially an interpretative framework that does not enable the formulation of interventions with directed theories of change, which is of our interest in realizing the purposive goals of doing social good.

ANT also does not provide any rules of thumb or generalizations of how various types of power differentials between the actants may influence the transformation of the network. Insights from systems thinking and cybernetics-based models are useful here. Benefits of systems thinking (seeing the system as a whole) to govern its behaviour towards the desired objectives, is discussed as being useful in the ICTs for development space (Turpin & Alexander, 2014). It helps to see technology in the wider context of social systems comprised of different kinds of actors who interact with one another. Useful concepts such as *open* and *closed* systems, and *boundaries* of a system, help determine the extent of complexity that is chosen to be modelled, and to consequently remain aware of what was not modelled and could result in surprises. Concepts such as *functions* that relate inputs with outputs of *flows* and *stocks* of various quantities, and composition of functions in dynamical systems that could lead to *emergent effects*, make precise the relationships that are modelled and assumptions therein. *Positive* and *negative* feedback loops are another useful construct to track and remain alert for emergent phenomenon. Decomposition of large complex systems into smaller hierarchically organized independent *sub-systems*, is also a useful technique to simplify the models. Finally, systems thinking and cybernetics allow the specification of concrete *objectives* that the system should achieve, such as for utility maximization or for system equilibrium. Modelling these *objectives* under different conditions and assumptions can help answer questions of whether to choose a particular system configuration or not, whether the system will find an equilibrium over time, or whether the regulatory loops for positive and negative feedback loops are strong enough to ensure system longevity. The models can be made as simple or complex as needed.

I build a modelling approach for power relationships on the basis of ANT, systems thinking, and the typology of power relationships. This modelling can be especially useful as part of the *system design* element introduced in Chapter 4 (Fig. 4.1), to legibly describe the power relationships within which the technology system is expected to operate and potentially alter, to meet the goals of bringing power-based equality of different types. As discussed in Chapter 5 though, envisioned goals are often not achieved by design alone, and need active management as well. The modelling approach I suggest can also be used to evolve management practices by tracking system and network transformations: it can guide managers to react appropriately by evolving policies or changing the underlying technology design, and it can also help them justify decisions or put them up for review by being able to legibly state the system model and the power dynamics it intends to shape.

6.4.2 Power Relationship Modelling

I propose to model power relationships for technology for social good projects in terms of the following constructs. As in ANT, individuals, groups, organizations, or technologies can be considered as *actors*. These actors may possess resources – a form of *power-to* relationships – for them to access resources. The resources could be about having the know-how required to execute certain activities, wealth to procure other resources, information required to make decisions, and discretionary or veto rights by way of official position or rank to make decisions. The conjunction of an actor and a resource becomes an *actant* that can bring about various actions. These actions I call *activities*, which can be of different types. An *activity* instantiated between a set of *actors* may result in changing the resource distribution among these *actors*. For example, redistribution of wealth takes wealth from the wealthy and distributes it among the poor based on some wealth distribution rules. An *activity* of knowledge sharing, on the other hand, would augment the knowledge resources of the recipient *actors* without resulting in any corresponding decrease in knowledge for other actors. The execution of an *activity* may further be controlled by *decision functions*. A *decision function* takes as input the resources possessed by an *actant*, to determine if the *actant* has the required resources to perform the *activity*. For example, an *actant* may be able to receive knowledge only if the *actant* has the time to participate in the learning exercise (time in this example is a resource) and the *decision function* would specify it as a pre-requisite to receiving knowledge. *Power-within* can also be considered as a resource that will be required for an *actant* to perform certain *activities* such as for a wife to have equal decision-making power as her husband over the financial expenses in a household. The decision function could also model *power-over* relationships, for example, wealth distribution may happen when an assessor values the current wealth levels of all *actants* but the assessor holds *power-over* the *actants* in making a correct assessment of their wealth. This represents one of the most common forms of *power-over* relationships, through decision control, where some *actants* may possess discretionary or veto rights to make decisions that influence others (Lukes, 2004). When a network of *actants* is stable as an entity in itself, it can be *black-boxed* into a single *actant* held together by *associational power*, and capable of imposing *structural power* by influencing decision functions for other activities. Such a configuration also models *power-with* relationships. The overall system can then be modelled over time, like a dynamical system, as activities are executed through the rules coded in the decision functions. Activities may also be guided through encoded norms, such as for rational behaviour, or mechanism and market design.

I demonstrate the modelling framework through a few examples. A promising future research direction may be to computationalize the model so that it can simulated through agent-based modelling to draw insights about the achievability of desired system goals.

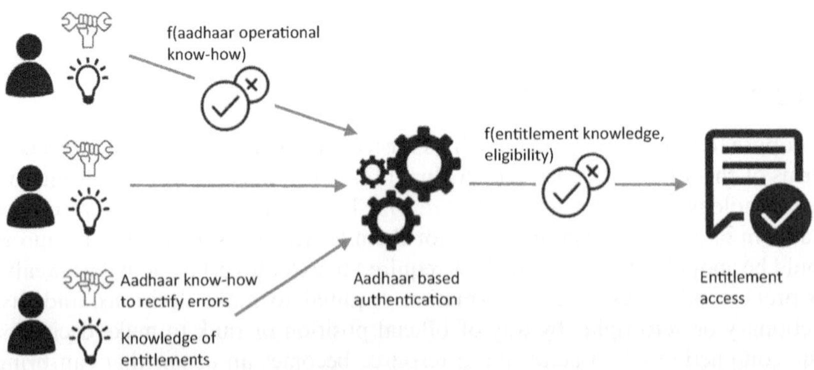

Fig. 6.1. Aadhaar-based Authentication as Controlling Access to Social Entitlements. To the left are shown beneficiaries seeking access to entitlements. An Aadhaar-based authentication system, interposed as an OPP, is used to control this access: only successfully authenticated users can get their entitlements. People need *first level* resources of knowledge about entitlements and know-how to rectify errors such as spelling mistakes in names or addresses in their Aadhaar enrolment, to have the *power-to* avail entitlements. Decision functions shown as f(..) reveal whether know-how and knowledge, and eligibility of people, will result in them being able to access their entitlements.

Aadhaar as an Authentication Platform. Fig. 6.1 shows a partial representation of an activity on the Aadhaar system of providing an authentication service for access to welfare schemes. There are three actors in this system: a beneficiary, the Aadhaar system, which accepts or denies beneficiary authentication, and the actual service availed by the beneficiary. At the *first level* beneficiaries must have the *power-to* possess resources such as the know-how or capability to engage with the Aadhaar system to operate it successfully. Those with less know-how may face problems such as rectification of registration errors or to demand the bypassing of Aadhaar in situations when network connectivity problems cause authentication failure. The activity of Aadhaar-based authentication is, therefore, governed by a decision function that is dependent upon the know-how of the beneficiary. The subsequent activity for availing the service is governed by the output of the prior Aadhaar authentication activity, and, of course, whether the beneficiary is entitled to the service.

Fig. 6.2 shows an alternative scenario where strong community relationships among the beneficiaries, perhaps intermediated by social workers and volunteers, leads to sharing of know-how, providing assistance to those who need it. This power gained through association *with* others helps people gain the *power-to* handle Aadhaar authentication failures because the beneficiaries are now better informed, and may get assistance from social workers and volunteers to help them work through the process. Clearly, inserting the Aadhaar technology by itself into the system and giving everybody equal access to it may not result in equal access to welfare schemes. Power-based inequality emerging from the possession of unequal *power-to* relationships between beneficiaries and their know-how cannot

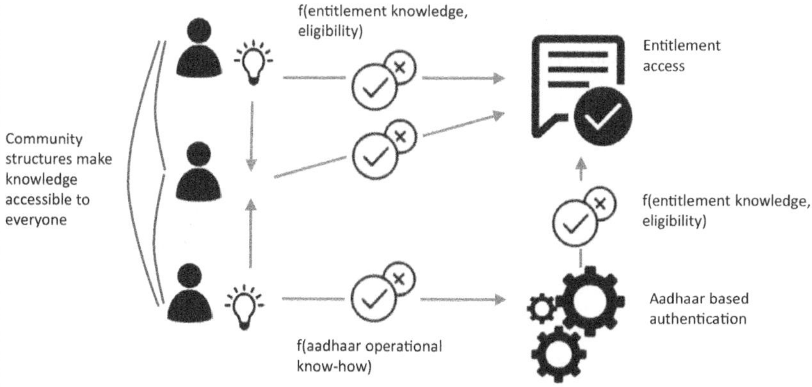

Fig. 6.2. Community-support to Handle Cases of Failures with Aadhaar-based Authentication. If beneficiaries do not have the requisite know-how to rectify issues with their Aadhaar registration, they can be denied access to their entitlements. Here is shown a scenario with two pathways: one, where community members are able to help one another acquire the knowledge or assistance to follow through with the Aadhaar processes, and two, a pathway without Aadhaar where beneficiaries can directly apply to access entitlements. These pathways counter the centralized position occupied by the Aadhaar OPP. Although not shown here, this also helps build institutional capacity in the community.

be ignored. Since the Aadhaar project does not emphasize mechanisms through which people can acquire an equal access to know-how or assistance to use the service, we can say that it does not hold power-based equality as a core value, which is essential for a social good project.

Access to Welfare Schemes. Fig. 6.3 shows a different scenario where access to a welfare scheme may be controlled by an official with discretionary rights – a *power-to* relationship at the *second level*. The decision function for the activity to access welfare in this case is dependent upon the eligibility of the beneficiary, and the discretionary rights exercised by the assessing official who thereby holds *power-over* the beneficiary. Access to grievance redressal services, or as shown in Fig. 6.4, a media service that can highlight cases of violation, can impose counter-*power-over* the local officials. The discretionary power commanded by them, which is held together through narratives of preventing leakages (*immutable mobiles* of ANT), is thus countered by alternate narratives of exploitation and domination highlighted by the media service. This activity of keeping a check on officials' discretion is therefore a regulatory loop. This also raises a question of quantification: how is this counter–power gained by the media service, and how large does it need to be to ensure that officials do not deny welfare services to deserving beneficiaries? The counter–power is likely to be controlled by resource variables such as the credibility of the media, its viewership or audience size, and whether the audience consists of higher officials to whom complaints about assessing officials can be reported. Many of these aspects are measurable directly or through proxy indicators, but will need a more complex network description.

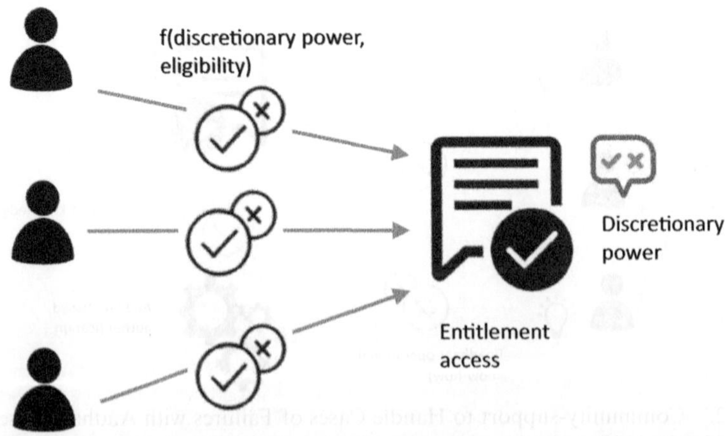

Fig. 6.3. Discretionary Power Vested in Administrators to Control Access to Welfare Schemes. Rules make discretionary power available to administrators that give them the *power-to* deny entitlements or grievance redressal to people. This is especially true when people do not have access to adequate legal or administrative escalation channels, or a good knowledge of the required processes and documentation, to protest against unfair denial of welfare. This then creates relationships where the local bureaucrats hold *power-over* the people to demand bribes for providing access to entitlements; a legitimate right.

A richer model is shown in Fig. 6.5 where the network has been expanded to include more actors such as the supervisors of assessing officials to whom reports of unfair denial can be escalated, and relationships such as the possibility of the media to extort funds from the assessing official to suppress the media report.

Role of the Media and Volunteers. Fig. 6.5 shows how the viewership of the media and its editorial credibility can be modelled. Additionally, community volunteers are also shown as actants who assist beneficiaries in submitting grievance reports to the media if the beneficiaries themselves are unable to do so, thereby bringing power-based equality in access *to* the media. The volunteers put pressure on the officials to respond to media exposure. Successful redressal and reduction in the unfair exercise of discretionary power, increases the credibility of both the media and the volunteers. Articulating the dynamical model also raises new questions about what the credibility function should look like, or how much minimum credibility should be attained for the media to become effective in exercising its influence. This is precisely the role served by models to identify places of oversimplification that need more detailing, which eventually leads towards a better understanding of the system. ANT indeed is able to provide a common language through which network structures and changes can be described, but the power-based model suggested here allows for a more formal description by articulating ANT devices such as OPPs, enrollment, and mobilization in terms of power relationships, and to evaluate outcomes by simulating different network configurations.

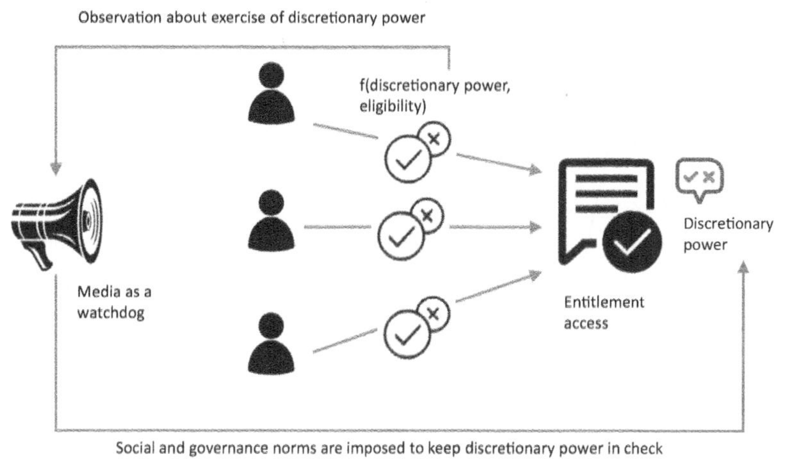

Fig. 6.4. Media to Keep a Check on Misuse of Discretionary Power. The media is an important institution to impose checks and balances on power holders. Stories carried in the media about the illegitimate exercise of discretionary power by administrators can counter–power-over relationships by putting pressure on the legal system or higher officials to react and address problems of rent seeking and unfair denial. This feedback function can prevent discretionary power from being exercised.

Social Media. Fig. 6.6 similarly demonstrates the need for regulatory feedback loops in a social media system. On the left is shown an open and unmoderated social media system where no checks or balances are imposed on posts made by the users. On the right, a reputation management system is introduced that uses the reactions to their posts to determine the reputation score for the user. This can be used to govern the exposure given to posts by this user, or whether the posts should be subject to a manual check by moderators. The nature of the regulatory feedback introduced by the reputation management system is important though, as explained earlier, because if it is biased towards metrics of engagement as on Facebook then it could amplify sensational news or misinformation instead of credible information (Tufekci, 2016).

Further, the reputation management algorithm can itself be considered as an actant that, by virtue of the rights vested in the algorithm, can impose *power-over* the social media platform users to determine how their posts are treated. Users can be given counter–*power-over* the algorithms by making them explainable and opening them up for appeals by the users. Similarly, persuasive user interfaces or norm shaping through social metrics can impose *power-over* the users to shape their behaviour and influence their decisions. Users can be given counter–*power-over* the user interfaces by making them configurable so that users can turn on or off different configurations.

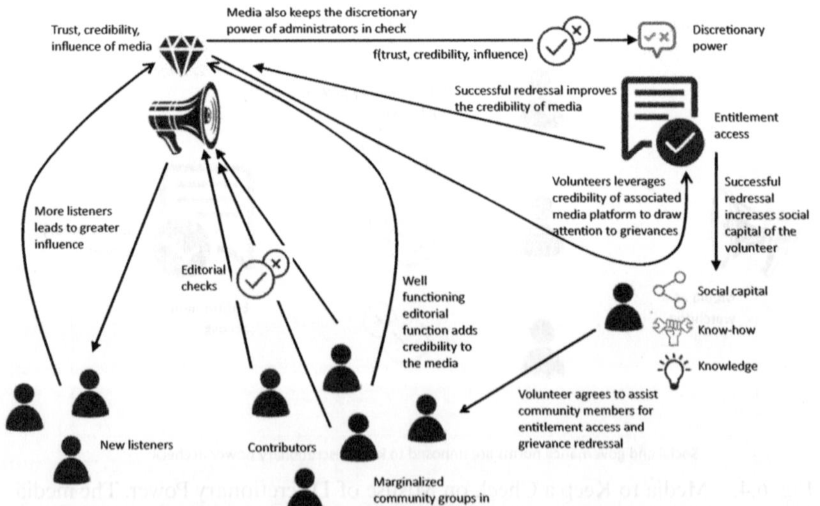

Fig. 6.5. Power Relationships Shaping and Shaped by Participatory Media Platforms. Shown here are three pathways through which the power of a participatory media platform may increase. Different pathways lead to strengthening of different types of resources. First, to the left is shown that the larger the audience of the media platform, the more influence it wields to create counter–power-over relationships. Second, the middle path shows that demonstrating a well-functioning editorial function will add to the credibility of the media platform, lead to greater community engagement, and consequently a larger audience size. Third, to the right is shown that facilitating grievance redressal will also add to the trust that people may place in the media platform. Similarly, volunteers who facilitate grievance redressal on behalf of marginalized communities will also gain social credibility as a resource. Feedback functions can thus be composed to model these dynamics. For example:

- influence ← number_of_users, that is, more the number of users, the greater the influence of the media platform;
- credibility ← accuracy(user_generated_content_selection), that is, correct decisions about accepting/rejecting user generated content will lead to greater credibility. Note that there may not be any universal notion of correctness of the decisions, rather it may change based on the community priorities of what kind of content they prefer, and would reflect the degree to which the media platform reflects the preferences of the community;
- trust ← trust + successful_grievance_redressal, that is, with each successful grievance redressal, the trust placed by the community in the media platform will increase;
- probability(exercise_of_discretionary_power) ← 1/counter_power(influence, credibility, trust), that is, greater counter–power gained by the media platform will reduce the chances of administrators using their discretionary power.

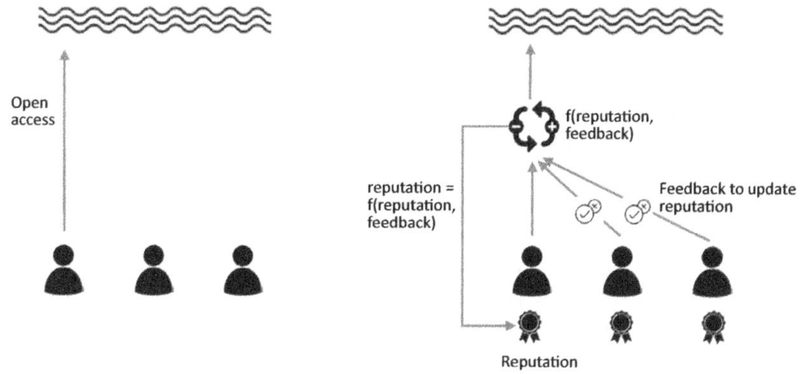

Fig. 6.6. Unrestricted and Reputation-based Publication in Social Media Platforms. On the left is shown a simple setup where anybody can access a communication channel on a social media platform to interact with their group. On the right is shown a setup where access to the communication channel is controlled by the reputation of users, modelled as a resource. The reputation resource for each user is shaped based on feedback received from other users on historical posts made by the user. The reputation is then re-calculated based on this feedback.

Modelling Procedure and System Archetypes. These examples demonstrate how the modelling framework can be used to study the transformations of power relationships over time. From a system design perspective, the model can be used to determine if a project has elements in place to bring power-based equality through counter–*power-over* relationships of checks and balances, as well as equal *power-to* relationships to not exclude deserving beneficiaries. From a management perspective, the network and resource transformations tracked by the model can be evaluated to determine if they are heading towards the desired power-based equality configuration or not. Visual maps can also be used to show temporal variations in the power configurations (Bengtsson & Lundstrom, 2013). Since these models encode social processes, they are unlikely to be very accurate for

These feedback functions can thus indicate whether the media platform can impose the checks and balances on administrators. This modelling exercise raises several questions about which variables should be modelled and how, such as whether credibility or trust is modelled better as a linear function or an exponentially increasing function, and what should be the form of the decision function to create counter–power. Reasoning about these dynamics can improve the chances of a participatory media platform meeting its purposive social good goals.

Fig. 6.5. (*Continued*)

prediction purposes, but the legibility they bring can help diagnose the current state (Solon Barocas et al., 2019).

I next outline the procedure to build such models, drawing attention to three system archetypes that often lead to predictable power effects. First, all actors that interact with the technology system should be listed. Second, the actors should be connected through different activities on which they engage with one another. This step is similar to the soft-systems modelling methodology of formulating *root definitions* of the interactions between different sub-system components (Torlak & Muceldili, 2014). Third, decision-making functions that govern these activities should be described. This requires an enumeration of power relationships that control the decision-making function, including relationships of access to resources by the actors. Methods of user inquiry and community workshops can be used to uncover these power relationships (Hunjan & Pettit, 2011). These relationships can also be informed through examples of other projects, or based on prior experiences in the concerned project itself.

Once the system is modelled, it can be analyzed for the presence of a few system archetypes. First, a star shaped or hub-and-spoke connectivity structure of the activity network is likely to contain actants that would be the loci from where power relationships emerge. Identification of such actants can help probe their power relationships deeper during user interactions. Second, a prior unequal distribution of resources across similar actants is likely to lead to further inequality of these or other resources. Resources which are unequally distributed from the outset should especially be examined in more detail during user interactions to understand their effect on decision functions, and rules can be introduced to distribute the resources more equitably. Third, the presence of regulatory loops can counter dominant power relationships. These loops can be examined for their effectiveness by conducting periodic user interactions to understand whether or not they are functional and having the desired effects.

The modelling approach I have proposed can therefore be used to compare different system designs with one another when conceptualizing a technology for social development project. The dynamical system formulation can then help test assumptions and evaluate outcomes as the project unfolds over time. Management practices can track the power dynamics and spot whether or not the relevant project goals along with the terminal goal of power-based equality are being met, and take appropriate corrective actions.

6.4.3 Technology as an Actant

Note also that I have so far applied this framework to understand power relationships at the socio-technical interface of different stakeholders who engage with a technology. The technology itself can however also be considered as an *actant* that can also impose *power-over* the users, such as reputation management algorithms on social media platforms, and I have briefly discussed methods through which users can be given counter–*power-over* this technology *actant*.

Another example I now consider is Internet of Things (IoT) projects that are projected to improve agricultural productivity, or big data-based approaches that use satellite data and other large datasets to make farming recommendations (Jayaraman et al., 2016; Woodard et al., 2016). Although these technologies are believed to be scale-neutral, it is feared that they would benefit only corporations and big farmers who can afford them, and thereby increase inequality (Fleming et al., 2018). The technologies could also result in de-skilling of farmers by spoon-feeding them with actionable information that would reduce the learning that farmers otherwise acquire through their own experimentation (Chaudhuri & Ghosh, 2020). This can increase the dependency of farmers on them. Even worse, these top-down techniques may not work well because they are often not be able to capture the vast complexity of ground realities, which can only be handled through experience and practical insight. Scott terms this practical knowledge as *metis*, and warns against top-down schemes that impose standardization, which eventually leads to their failure and hurts the people it was meant to serve (Scott, 1998).

If system boundaries are further expanded to also include the technology providers as actants then new kinds of power relationships may become visible. For example, providers can misuse the farmer data they acquire through their platforms, by providing data access to other actors such as traders or insurance providers. This can lead to negative power relationships, if the traders are able to use this data to bargain for lower prices with the farmers. Forms of counter–power can emerge if users are compensated in some manner for their data, so that they can define what the data gets used for, and regulations are imposed to audit the claimed efficacy of these technologies to benefit the farmers, or to even completely reconsider the use of technologies that make users dependent upon them and open new spaces for their exploitation. These are precisely some of the concerns that civil society has raised regarding recent proposals in India on the Agri-stack (GoI, 2020a, 2020c).

A similar argument is made by Schumacher (1973) who advocates the use appropriate technologies that can be controlled by their users. He advocates to not use external technologies that are manufactured or managed by large corporations, and to rather use small technologies that can be locally built and managed. He goes on to argue that external technologies are actually harmful for developing regions because they do not build skills of the local people and do not have regard for the local environment. Comparisons can be drawn with the technologies discussed above, which, on the one hand claim that they are neutral platforms on which users can freely participate and innovate, but on the other hand also impose narrowly defined interfaces that do not provide sufficient room for the users to steer the platforms according to their needs. This introduces new forms of *power-over* the users with little room for the users to impose any counter–*power-over* them.

6.4.4 Other Frameworks for Power Modelling

The Powercube modelling framework is quite popular in the social development space (Gaventa, 2019). It characterizes power analysis along three dimensions: *levels* (global, national, local, and household) at which relevant power relationships should to be considered; *spaces* (closed, invited, and claimed) to understand the type of access that stakeholders have to decision-making; and *forms* (invisible, hidden, and visible) of the power relationships being observed. This characterization is useful to conduct community-reflection workshops through which community members can understand and make legible the kind of power dynamics they experience. Goals and strategies can then be developed to alter these dynamics. The modelling framework I have discussed is complementary to Powercube: it helps analyze power relationships between actors and how these relationships are manifested through activities, whereas Powercube helps delve deeper into decision-making processes, including those related to the political economy of the state and markets. My proposed framework can be expanded to incorporate the Powercube vocabulary, especially to characterize the system boundaries that have been considered in the model.

Another tool for power analysis is Netmap (Schiffer, 2006). It uncovers power relationships through community discussions by identifying outcomes that emerge as a result of domination by some actors, and thereby makes legible to community members the actors and sites of concentration of power. The relationships uncovered and modelled in Netmap are similar to *power-over* relationships. It does not model network relationships though, or power that arises from the possession of different resources.

The view of innovations as leading to transformations in power dynamics has recently been considered in a new theory called the transformative social innovation theory (Haxeltine et al., 2016). The framework I have proposed gives a concrete methodology to this theory, to ensure that unjust disempowerment of the weak does not happen, and that power relationships between different stakeholders are shaped towards equality.

6.4 Summary

I have discussed so far that meta-social good projects are those that facilitate society to discover values it considers important, and to address gaps in the mechanisms of the state and markets that may side-line these values. I also discussed how power relationships of exploitation or domination are behind the inability of social systems to meet these goals, and to entrench the current systems more strongly. Power-based equality therefore needs to be a core value for all transformational meta-social good projects. Since meta-social good projects define the values essential for social good, power-based equality should also be a core value for social good projects to uphold. I described several social good domains to show that understanding power dynamics and shaping them is at the heart of doing social good. Technology for social good projects should therefore be viewed as meant to intermediate social relationships between designers and users, or managers and workers, or between different users themselves, or between technologies and users, such that users are not disempowered by the introduction of technology. Creating relationships of domination that disempower the weak is incompatible with humanism and social good.

Being able to alter power dynamics that are entrenched in the world today amounts to taking a political side. Picking a political side is, however, not easy. It is not easy for technologists who are typically trained in the rationalist tradition of science that specifically claims to be politically neutral (Escobar, 1995; J. Moore, 2020). It is not easy for consumers or citizens either, given that they are influenced by media messaging and the wider political economy of technology (D. Harvey, 2003; J. Pal, 2008; A. Sen et al., 2019). In the forthcoming chapters, I outline these challenges in taking a political side. I suggest mechanisms through which technologists can play a stronger political role in holding their organizations and governments more accountable towards ethical principles (Cooley & O'Grady, 2016; Seth, 2021a). I similarly outline the cultural hegemony within which political and technology debates take place that impedes having more and more people adopt technologies genuinely meant for social good, and I outline mechanisms through educational curricula and communication platforms that can empower people to overcome this hegemony (Gramsci, 1971). This can be a pathway to impose more meaningful social control over technology so that it does not disempower the weak.

Chapter 7

Constraining Structures and Ideologies

In the discussion so far, I have advocated for several criteria to judge whether a technology project can be considered as a social good project or not. I have explained why incorporating ethics in the technology design alone is not sufficient to ensure that harmful unintended outcomes do not arise; management practices are essential as well. I have shown that these design elements and management practices also need to be consistent with one another on their respective ethics-based foundations. I have highlighted that social good projects should have some essential features that can distinguish them from other projects, and I stated that these should include having a consequentialist terminal stance and having a focus on achieving power-based equality as a goal. Projects that do not meet the power-based equality criteria run the risk of not being able to solve the social problems they aimed to address, and may disempower the people they intended to support. Projects that do not have any terminal values, even if they do uphold some instrumental values as guardrails, can be hard to steer away from harmful outcomes.

What does this mean for technologists – the designers and managers of the technology systems – who I believe would not want their labour to lead to harmful outcomes, and ideally to contribute towards social good by meeting genuine use-values for society and to empower the weak? Are they interested at all in contributing to political projects that can transform entrenched power relationships in society? Marx has discussed that the alienation that arises from a distancing of workers from the outcome of their work is ultimately dehumanizing and unsustainable (K. Marx, 1844). The desired state for all technologists would be to be more responsible and contribute towards positive social relationships that are fostered through the output of their labour. While employees at Google have protested against the use of Google's products by the military, and employees at Facebook have stood up for stronger internal regulation of algorithms and processes to prevent misinformation, such demonstrations are, however, rare when compared with the vast workforce of technologists in the world (Issac, 2019; Shane & Wakabayashi, 2018; Thompson & Vogelstein, 2018). In fact, many technologists knowingly and unknowingly continue to push for technologies that create disempowerment at different levels. What can explain this behaviour? I next

Technology and (Dis)Empowerment: A Call to Technologists, 131–144

Published under exclusive license by Emerald Publishing Limited
doi:10.1108/978-1-80382-393-520221007

highlight some plausible reasons for suppressing a resistance among technologists to acquire control and influence meaningful social relationships through their work. This suppression may in fact be intensifying an impending alienation among technologists, and in the next chapter I outline several steps that they can follow to play a stronger role in influencing their work, their organizations, and their governments, to build and deploy technologies for social good that remove hegemonic power structures in society. This may help identify a new utopia of work that is non-alienating and more humanist for technologists than the status quo.

7.1 Structures and Ideologies That Constrain Technologists

7.1.1 Differentiated Values and Priorities

A separation of concerns arises from the functional segregation at most workplaces that are structured on the basis of a division of labour. Engineers are trained to care for values such as correctness and cost efficiency, and not whether the systems they are building are inclusive or aimed at particular political outcomes (Riley, 2008). Designers are expected to bring in these values, and translate them into engineering requirements, but only if the values are deemed important by the organization itself (Duboc et al., 2020). This creates a distributed accountability among different organizational units. Technologists may even use this to justify a moral buffer that shields them from taking overall responsibility. This same reasoning has further prompted views on the neutrality of science and technology, especially among technologists situated upstream in the production value chain where technologies they build do not directly touch end-users. As I have however discussed in Chapter 3, this view is not only incorrect, there are many examples where scientists acknowledged their responsibility irrespective of their location in the production value chain and challenged the misplaced importance given to functional segregation (Berg, 2008; Russell Einstein Manifesto, 1955; Wiener, 1950).

Further, the rationalist approach in most engineering and design methods gives the illusion that systems designed with particular values uphold these values post deployment (Riley, 2008). However, as I have discussed earlier, values-based management of technology is also essential, and requires a close connect between engineers, designers, and users (Seth, 2019b, 2021a). Bo Dahlbom and Lars Mathiassen discuss this in their analysis of a romantic versus mechanistic view of information systems, and argue that the outcomes of technological systems cannot be mechanistically determined (Bo Dahlbom & Lars Mathiassen, 2016). Rather, the outcomes are shaped by the context in which the technological systems are deployed. The culture, power relationships between the users, and the agenda of users behind using the technology, all influence the eventual outcomes, and are in turn shaped by the technology as well. Facebook's claim of capturing the complexity of society in legible algorithmically driven community standards, or the arrogance of Aadhaar's designers in underestimating the need for protocols to handle cases of technology failure are examples of a mechanistic – rather than romantic – view in approaching problems. A belief of doing social good based on what may be a fundamentally mistaken assumption that the expected goals will be inevitably achieved through pre-determined pathways may even reduce the motivation for technologists to try and deeply understand the real outcomes emerging from their work.

7.1.2 Fragmented Organizational Structures

Even if a romantic view were embraced by some technologists, the fragmented organizational structures and distributed technology production value chains prevalent in today's industrial society can impede the free flow of information (Suchman, 2002). Engineers, for instance, may not hear what ethnographic study teams have to say about problems that some users segments may be facing, and these problems may therefore go unaddressed. Further, different teams and individuals may be operating under different ethical systems. Insufficient socializing among them can lead to inconsistencies in how they choose to respond to various observations. Teams that interact with users in person may be more empathetic in their response to user problems, whereas teams that are removed from direct user interactions may choose to prioritize other issues, or respond differently. In fact, physical segregation between different teams could even be deliberately imposed by having different offices, such as between business teams and engineering teams, to prevent association and maintain ambiguity in organizational values so that it becomes harder for undesirable values to be pinned down and challenged (Prado, 2018).

It may also not be straightforward to have an organization-wide ethical approach incorporated in the day-to-day actions of different teams. This is clearly seen in the case of data privacy, where even a unanimous embrace of strong privacy norms by companies by designating Chief Privacy Officers and setting up a dedicated privacy department, has essentially not moved beyond serving a compliance function so that privacy rights continue to be exercised through the inadequate notice and choice mechanism (Waldman, 2018). The power of the specific departments within companies, their integration with different teams, resources available for education, and sensitization of large teams to concerns raised by the department are common reasons why the adoption of a uniform set of underlying ethical principles remains broken within organizations (Pfeffer, 1992). Silos, therefore, continue to prevail, and can distance technologists from developing a motivation to understand the outcomes from technologies built and managed by them.

7.1.3 Political Economy of Technology

The wider political economy of technology serves to entrench naive optimism about the role of technology in society. This happens through several steps.

Fig. 7.1 shows the business and political landscape within which technologists are embedded. Shown on the left are factors that influence the technologies designed and managed by private enterprises. The nature of most technology used

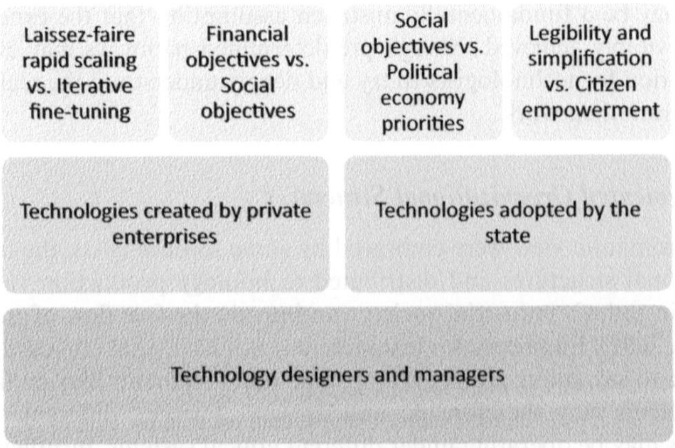

Fig. 7.1. Political Economy in Which Technologists Exist.

in the information systems of today requires large amounts of capital, which in turn requires financial investment to put the company on a path to rapid scale-up to meet investor expectations for quick returns on capital, as opposed to growing slowly with careful iterations of the design and management processes. Scale is thus made into a goal in itself, and metrics such as users per employee and profit per employee are then used to justify to technologists the impact that their technologies are having (Bryan, 2007). Instead of upholding values such as power-based equality for social good, or outcome metrics that instrument the actual impact that a technology project may be having on people, an ideology is created of measuring success in terms of the number of users irrespective of the outcomes emerging from the usage. Technologists fall for such value-less metrics instead of focussing on meaningful impact. Monetizeable employee stock options may further align the interest of technologists with those of investors, towards scaling and market value rather than social good.

While rapid growth and massive influence on the lives of millions of users should emphasize the role of responsibility even more, on the contrary, technology companies resort to extensive lobbying to avoid regulations and restrict their liability in the case of harmful outcomes (Kang & Vogel, 2019; Zuboff, 2018). A fallback on arguments that the harms were truly *unintended outcomes* is often used by organizations to evade taking responsibility (Nohria & Taneja, 2021). The practice of ethics is delegated to technology design, and instead of legally ensuring even this limited view, ethics is often left to self-regulation (Wagner, 2018). Interestingly, responsibility is also placed upon individual technologists to be more careful in their work, even while they may not be encouraged to connect with or to understand the outcomes of their technologies (Stark & Hoffmann, 2019). Ethics review procedures for research at universities are strong, with the view to put forethought in considering possible harmful outcomes of the research, but similar reviews are not required when companies build and scale new innovations that can affect millions or billions of users. Medical ethics has such procedures but information technology has managed to keep itself outside the purview by arguing that any regulation can hamper valuable innovation (Wilemme, 2018). This manipulation may thus effectively destroy the emergence of responsibility and accountability: on the one hand, technologists are advised to be more responsible, on the other hand companies evade taking accountability and may not be truly committed internally to adopt ethical norms. This orchestrated ambiguity thus dampens technologists' motivation to critically examine the technologies that they build and operate.

Under these conditions of taking pride in growth statistics, which is actually driven entirely by financial goals of investors, propaganda is further created to emphasize the benefits of technology instead of its harms, and strengthens the belief among technologists that they indeed are contributing towards social good. Gillespie demonstrates how technology platforms present themselves as neutral sites to facilitate transactions and innovations while in reality it is just a tactic to evade regulation (Gillespie, 2010). Gig-economy platforms especially present themselves as benefitting workers by giving them the flexibility of working hours and bringing them more opportunities, while they may in fact be manipulating and exploiting the workers (Dhillon, 2018; Rosenblat & Stark, 2016). The inconsistencies demonstrated between the public statements made by gig-economy companies, and their actual practices, has been termed as *fair washing*, to emphasize how these companies falsely attempt to portray themselves as supporting the welfare of workers (Katta et al., 2020). Fuchs similarly examines social media, of how it effectively extracts value from the unpaid labour of social media users, glorifies simplistic notions of participation instead of nurturing understanding and cooperation among users, trivializes complex discourse into small fragments, and entrenches existing power relationships through its advertising business model to influence the behaviour of people (Rivera, 2020). Social media companies, on the other hand, highlight the communication and empowering effects that their products create (Vaidhyanathan, 2019; Zuckerberg, 2019).

While corporations may not be genuinely interested in doing social good, and may even attempt to co-opt the concept, shown to the right of Fig. 7.1 is that even governments and social enterprises often follow opportunistic goals and create an ideology of *high modernism* about their work. Governments tend to use technology as a means of imposing greater coordination and control of the population, but market these goals as social good even though many such initiatives have often disempowered people and reinforced inequities (Scott, 1998). As examples of Aadhaar-led digital approaches to social entitlements in India have shown, politicians may present technology as a novel solution to social problems, while the shortcomings are obfuscated, and the media further creates an aspirational appeal for more technology, even though the technology design and management may not be aligned with social good (R. Khera, 2018; J. Pal, 2008; A. Sharma et al., 2020). Further, companies are in constant search for customers for new technologies, and ingeniously market new solutions to governments for their own benefit (A. Sen et al., 2019). Capitalism needs continuous technology innovation to survive, and hence neoliberal capitalism cleverly makes the media and state complicit in manufacturing the need for innovation (D. Harvey, 2003). Tight interlocks between governments and corporates through networks of the power elite, and the ability of companies to influence policy due to growing inequality that places the power of capital in their hands, enables them to shift the state's priorities away from genuine social objectives to what solely benefits the corporations instead (C. W. Mills, 1956; A. Sen et al., 2018; J. E. Stiglitz, 2012). Aadhaar and various technological layers built on it (collectively called the India Stack) are examples of how corporate and government interests have aligned to create an entire entrepreneurial industry that claims to be doing social good, even though realizing the stated benefits has many unaddressed gaps (Dattani, 2019). Zuboff similarly shows how the surveillance capabilities of technologies produced by large corporations found resonance with the social control needs of governments after the 9/11 incident, and dampened rather than intensified the emergence of regulatory mechanisms to keep a check on inappropriate use of data (Zuboff, 2018).

Ideologies of high modernism are also encouraged at the global level through institutions such as the World Bank. The Washington consensus projected development as a goal, and that it could be managed through professionalism and expertise imported from developed countries into the third world (Escobar, 1995). This, however, can lead to an increased dependency of developing countries on external technologies, and ignoring local social structures has led to a failure of many such externally imposed projects (De et al., 2018; Urquhart & Andrade, 2012; Wade, 2002). Mechanisms such as hackathons and startup toolkits, however, continue to be regularly used to identify new use-cases in the Global South for technological concepts invented in the Global North (Avle et al., 2017). All this ends up reinforcing the technology optimism and high modernism ideology of technology among technologists, educational institutions, governments, and society.

7.1.4 Regulatory Gaps

Lessig highlights deficiencies in the law-making process to address policy vacuums that are invariably created with technological change (Lessig, 1999). He takes the example of the Internet, which introduces *latent ambiguities* in aspects such as copyright over intellectual property, people's rights to privacy, and the governance of free speech, compared to how they were handled in the pre-Internet era and what changes in law may now be required. He first shows that the Internet is governed by its architecture (or code), and therefore law to handle ambiguities must deal with the governance of the Internet architecture. He then shows that new laws have failed to do so successfully because the architectural changes advocated by these laws have often violated fundamental values which were respected in the pre-Internet era. For example, on the aspect of intellectual property, the law has sided with giving perfect control to content creators to track distribution and re-distribution of their content through technologies like Digital Rights Management (DRM) and laws like Digital Millennium Copyright Act (DMCA), which conflicts with fair-use principles to limit copyright for wider cultural reproduction. Similarly, on the aspect of privacy, the law has primarily evolved notice and choice mechanisms by publishing terms of service, which however, is an ineffective method. Lessig claims that better solutions are possible through a better mix of law, social norms, markets, and code. On the aspect of free speech, the law in many jurisdictions is tilting towards regulations that make censorship of communication easier, but which makes even legitimate democratic dissent harder. Lessig argues that law has often not been able to find good solutions because the practice of law does not understand technology closely enough to realize that technology architecture embodies values, and that any changes in the architecture should be evaluated in terms of these values. Mistakes can prove expensive by altering the hard-earned balance of rights in society that came about through centuries of thought and activism. Further, society itself does not understand this well to be able to shape law directly instead of letting it be shaped through the so called policy experts, where chances of corruption due to market or political influence are much higher.

Nissenbaum takes this further by explaining how laws for technology regulation come about (Nissenbaum, 2011). She observes, similar to arguments put forth in the earlier chapters, that technology design alone cannot galvanize technology from harmful outcomes. When these outcomes differ from values considered important by society, either society adjusts by rejecting the technology or learning to use it safely, or laws and procedures are put into place such as for appeals, liability payments, and safety assurances. Which of these directions is chosen is, however, a political outcome, influenced by lobbying, protests, media coverage, and other political economy factors. Thus, a layered structured can be conceptualized, of some values ensured by the technology itself, some values enshrined in laws related to the technology production and use, and some left to the ethics of users or of technologists who build and manage the technologies. Technologists therefore cannot outsource their morality to technology or legal institutions because gaps will inevitably remain in the extent to which the technology or regulations can ensure a complete adherence to the values agreed upon by society.

7.1.5 Summary

In summary, I believe that technologists cannot avoid their crucial role in preventing disempowerment of the weak arising from their work. However, technologists are systematically distanced from understanding these outcomes through a combination of educational conditioning, fragmented organizational structures and distributed technology production value chains, reinforcement of a dominant ideology of positive and deterministic outcomes from technology, and lack of acknowledgement of the need to improve technology regulation. Just as the *false consciousness* of how factory workers wilfully embody the ideology of the ruling class, in the same way educational institutions, workplaces, companies, and governments, perpetuate ideologies of high modernism among technologists (Little, no date). Although most practitioners and technologists may be well-intentioned, they are trapped within this narrow ideological frame through which they have been trained to look at the world and examine their work. The blind spots of harmful outcomes thus go unquestioned and the concept of social good is further co-opted to create a society-wide cultural hegemony of a positive and infallible character of technology. These misleading ideologies prevent technologists from committing towards the political project of doing genuine social good, meant to empower the weak through a transformation of unequal power relationships in society. In the absence of this resistance, technology developed within the current paradigm tends to reproduce existing systems that exploit the weak and increase inequality.

7.2 Alienation and Resistance

Misleading ideologies may not go unchallenged for long though. Deliberately influenced through such misrepresentations about the outcomes of technologies, and being unable to influence these outcomes, can be considered as labour control of technologists. This makes it hardly different from the labour control experienced by blue-collar factory workers, and according to Marx, this creates alienation which sooner or later will give rise to a resistance by the technologists themselves. I next discuss Marx's concepts of humanism and alienation in more detail, and if they indeed apply to technologists.

Several different concepts of humanism have been articulated in the past (Hodges, 1965). During Aristotelian times, humanism was equated with self-realization and excellence, implying an all-round development of personality. Further, it was not meant to be accessible to everybody, only to the aristocratic class who had the capacity and freedom for intellectual thought. John Stuart Mill developed a very different concept of humanism as linked to the welfare of others. A need for people to identify themselves with the happiness of others was stated as being an essential component of civilized life. John Dewey rejected the difference between intellectual and manual labour, equating both of them as problem-solving, and developed humanism as intelligent problem-solving that was aimed at human growth. This growth included many dimensions, such as becoming educated, in favour of ordered change instead of violence, and accumulation instead of destruction. Marx's concept of humanism, which I have used in this book, subsumes many of these concepts. Similar to Dewey, Marx did not distinguish between intellectual and manual labour. Like Dewey and Mill, Marx recognized humanism as fulfilling social needs. However, the process for fulfilling these needs was stated by Marx as being through relationships of production – humans could be of service to others only if they produced outputs of use-value through their labour; and humanism lay in building positive social relationships between producers and consumers through this produced output. This concept of enmeshing humans into a social network of production–consumption relationships as humanism, was starkly different from the concept of Mills' as welfare, and of Dewey's as growth. Marx also saw the production process as being built on social relationships when multiple workers came together to produce an output, and built solidarity through team work. Labour was therefore seen as a social process, both in production as well as the distribution of its outputs. Consequently, social relationships that were created coercively in the production process such as between capitalists and workers, or in producing outputs for contrived use-values such as between marketers or advertisers and consumers, were considered anti-social.

Marx further saw this as the origin of private property. When humans see themselves as a commodity that can be sold without necessarily leading to positive social relationships, they convert their private bodies into a property that can be traded, acquired, accumulated, and used as an instrument of labour with no obligations to foster positive social relationships. The loss to them from what should have been social labour now turned anti-social, appears as surplus value which becomes profit for capitalists. The loss of humanism thus forms the basis of Marx's theories about capitalism as devouring humanism in society and converting it into capital which is accumulated by a few. The de-humanizing alienation of workers thus arises in different forms – in the production process, from the output of the production process, and the outcomes in society affected by the produced goods and services – and leads to profits for capitalists. This makes the relationship between capitalists and workers fundamentally exploitative and anti-humanist. Gorz in his book *Critique of Economic Reason*, similarly explains how economic rationality is at the heart of this anti-humanism (Gorz, 1998). Economic rationality reduces labour to a tradable commodity measured through a single dimensional value, and this quantification then makes it possible to divide, combine, and substitute various factors to find the most cost effective and optimal combination. Imposing this rationality however requires bureaucratic structures to regulate behaviour, such as in coordinating production processes according to Taylorist methods, ensuring free trade and exchange for competition in markets, and honouring contracts between capitalists and workers engaged in a production process. Such regulation becomes dominating and takes the nature of work away from voluntary labour of love and care that fosters positive social relationships, to sustaining coercive relationships between capitalists and workers, divisive relationships between the workers themselves, and instrumental relationships with society members who are nudged towards consumerism to fulfil pointless needs. These bureaucratic structures of domination are thus needed by capitalism to maintain the hegemony of economic rationality, while also operating within its rules. Humans then compete and pit themselves against one another in a spiral of increasing exploitation and alienation, de-humanizing themselves and converting themselves into commodities.

A praxis of resistance was seen by Marx as being the only method to overturn this exploitation (C. Fuchs, 2021a). Marx rejected hegemonic structures of domination and accumulation that could permanently control people, he believed that the alienation that results from exploitative production relations will eventually lead to collective action by the workers to overthrow these structures. Such action will however first require workers to overcome the false consciousness of individualization that persuades them to compete with one another in selling their skills and time for wages. Marx, and later followers like Antonio Gramsci and Paulo Freire, strongly believed that all workers are intellectuals in themselves and the praxis of resistance needs workers to bring theory and action together. Methods like workers' inquiry that encourage workers to discuss and share experiences, are believed to be essential for this process (Haider & Mohandesi, 2013). Further, one of the dimensions of capitalism to continuously innovate on technology so as to replace or control labour, was also believed by Marx to lead to an improved skilling and education of the workers. He believed that as workers become more educated and aware, it will become easier for them to coordinate with one another to overcome the hegemony of capitalism (Adler, 1990).

Foucault similarly states that the use of corporate and state power to perpetuate ideologies through the media, workplaces, and schools, shapes the forms of knowledge that society considers legitimate, and knowledge in turn shapes the ethics of society (Foucault, 1984). This hegemony can only be overcome by countering the knowledge with other forms of knowledge, and is the key challenge for a praxis of resistance. Foucault further believed that such knowledge can emerge from local struggles, in invisible and hidden spaces, against the established structures of power (Lukes, 2004).

Workers' Inquiry

The workers' inquiry method is based on the premise that only the working class can provide meaningful information about its own circumstances of existence (Haider & Mohandesi, 2013). However, the goal of workers' inquiry is not restricted to just a self-ethnography, but that in the process of writing and discussing about their lives, workers would come to see that their problems are shared by many others. Worker newspapers carrying narratives mixed with revolutionary theory are meant to build such consciousness as groundwork for eventual collective action. With the same motivation, Marx had put forth a large questionnaire of 101 questions in 1880 in France (K. Marx, 1880). The questions were meant to directly connect the daily experiences of workers with the wider concepts of exploitation and surplus value appropriation by capitalists, so that workers can become aware about their predicament in a capitalist society and organize to arrive at a new system. As a side note, the same questions are interestingly just as applicable in current times among workers in India and the Global South more broadly, especially about working conditions and wage rates, showing that capitalism over the years has exported the inhuman working conditions of the nineteenth century West while it was industrializing, to the Global South now.

Gramsci similarly recognized the underlying problem of the hegemony of capitalism as arising from its ideology having been accepted by people as *common sense* (Gramsci, 1971). Countering it required the emergence of what he called *organic intellectuals* from within social groups to organize and direct the ideas and aspirations of their groups. They were different from *traditional intellectuals* such as scientists and professionals. Organic intellectuals specifically helped organize people through the awareness they had of the economic, social, and political domains of their respective social groups. Gramsci further considered the role of political parties to weld together the organic intellectuals and traditional intellectuals, and that this could happen if a common *active education* programme was followed in schools. This part of the school curriculum Gramsci envisioned would focus on the development of humanism as emerging from social relationships and collective life, before the students went on for further professional training. It would build a certain level of maturity, capacity for intellectual and practical creativity, and of autonomy in orientation and initiative. Professional work then would come to embody this essential humanism.

Freire brought similar ideas to his notion of *humanists* who would work alongside the oppressed and engage in their struggle, not out of generosity, but out of respect and trust for them (Freire, 1970). These humanists would, through a dialogic process with the people, support them in arriving at a common understanding about the world. Freire's emphasis was that an educational or organizational model that treated people as empty buckets that needed to be filled with knowledge would distance people from their own decision-making as if they were mere objects. It was therefore important for him that through the dialogic process the oppressed would begin to understand the exploitative logic of the oppressor. They would then reject a desire to emulate the oppressors as a marker of success, and instead build strategies to counter common techniques used by the oppressors to defeat revolutions. They would see through these common techniques such as divide and rule, manipulation, and cultural invasion to internalize inferiority, and eventually build a synthesis about the realization of oppression around which people can organize.

These approaches of workers' inquiry in the contemporary context have taken interesting forms by utilizing participatory communication tools. This has been heavily documented especially among gig-economy workers in delivery and ride-sharing services who actively use tools like Whatsapp to understand the opaque black-box of algorithmic decision-making that controls their work, and to then collectively overcome exploitative or dominating steps dictated by the algorithms through simultaneous log-outs or other mechanisms (Woodcock, 2021). Many other workers, however, may be driven to accept their conditions as inevitable and unchangeable, and not participate in workers' inquiry or collective action efforts. Worker newspapers have often grappled with the challenge that many workers are not interested to write, and describe their experiences as "not interesting" or that it "doesn't really matter" (Castoriadis, 1988). Our experience at Gram Vaani with running a voice-based platform for industrial sector workers employed in the garments and automotive industry in the Delhi region, is somewhat similar. Few workers contribute voice reports, and those who do are concerned about repercussions and prefer to either not openly talk about some sensitive issues or contribute through pseudonyms (Seth et al., 2020). However, stories of worker counter-power of strategies adopted by groups of workers in different factories and industries to secure their rights, are also actively shared, discussed, and similar actions are widely replicated through the network.

The workers' inquiry method is applicable even for technologists, to develop a critical understanding of the political economy in which they are embedded, and the organizational structures and ideologies which control their perceptions about work. To build a praxis of resistance against these disempowering technology paradigms, technologists will have a lot to learn from one another to introspect about their work, work environments, and organizational policies and priorities, that can guide towards ideas for action.

Is there any evidence that technologists indeed feel alienated when their labour is controlled through ideological misrepresentations, and will they participate in a resistance to change this?

There are few studies regarding the alienation of technologists and how they perceive their role in society or evaluate the worth of their work. A welcome addition is the recent book by Mike Healy, *Marx and Digital Machines: Alienation, Technology, Capitalism*, about software engineers, managers, and scholars working in the Information Technology (IT) industry (Healy, 2020). Interviews by Healy of engineers in services companies shows that it is indeed alienating when they are made to focus on atomized sub-tasks, which are assigned authoritatively by a manager, and with no insights to understand the end output. In fact, as many facets of the managerial function are being increasingly automated to increase developer productivity, this may even situate white-collar technology workers under algorithmic control, not very different from the control imposed on blue-collar workers (Asthana et al., 2019; Rosenblat & Stark, 2016). Further, it may even be a first step towards automating the job of a manager (Cooley & O'Grady, 2016). Healy found that even among elite technologists who may be doing more creative work, including academic researchers, there is an inherent desire to do social good and avoid harms, but they are unable to move out from their established work patterns since their continued incentives, performance metrics, and remuneration are aligned with the capitalist work structure which, therefore, they do not want to disturb. Thus, it seems that moral buffers or a disinterest in understanding the eventual outcomes of their work may actually be closer to a Freudian defence mechanism rather than an inherent characteristic of technologists. Possibly this is also why we are seeing an increasing backlash against technology by its users, many governments, and employees, alike. Alienation, as pointed out by Marx, is anti-humanistic and it is only natural that eventually a resistance will develop.

There is a rich history of such resistance movements. Norbert Wiener was among the first ones to draw attention to the need for technologists to control the purposes to which their technologies were applied. This was famously highlighted in his open letter titled *A Scientist Rebels* where he refused to share details of his technology design with militarists (Wiener, 1950). He went further to illustrate how totalitarian governments or profit-seeking capitalists ignore fundamental human values in driving the adoption and use of technology, and asked scientists to not be naïve and take responsibility so that their inventions are not used by others for unethical private or political gain. The Whole Earth Catalog started in 1968 had similar aims to critique products on the violation of values of openness, knowledge sharing, ecology, and appropriate technology (Wikipedia, 2022). Related ideologies such as the *hacker ethic* aimed at making computers accessible to everybody, to keep information free, mistrust authority, and to use computers for the betterment of society (Levy, 1984). The Free Software movement launched in 1983, and the cypherpunk movement started a few years later to use privacy enhancing technologies for social and political change, can be identified as resistances put up by technologists themselves against closed innovation, access restrictions, and societal control (Rogaway, 2015; Stallman, 2007). The Wikipedia itself is a foremost example of volunteers coming together to make knowledge more easily accessible to everybody. I discuss such movements in more detail in the next chapter.

Technologists have a lot to lose if they do not reinvigorate the deeper values behind similar praxis of resistance. First, there is a trend towards the use of information systems for authoritarianism and walled gardens of knowledge, both by governments and by companies (Reetika Khera, 2019; Zuboff, 2018). However, the science and engineering professions are founded upon unhindered creativity and the free movement of knowledge (Wiener, 1950). Authoritarian use of technology and closed knowledge structures restrict and corrupt the information flows needed for science and engineering, reducing both the ability of technologists to innovate as well as the effectiveness of their innovations. Second, even though technology can be leveraged effectively for genuine social good, the misrepresentation by corporations and governments to not highlight the challenges in doing so, is aimed at using technologists to further their own interests instead of having them solve social problems. Technologists are thus not only being exploited and converted into instruments for the ends of others, they are also bringing about their own alienation by contributing not to positive social relationships, but to those that disempower the weak. Third, introduction of automated tools to manage the technology development process itself, can impede the freedom of technologists. These trends point to an increasing alienation among technologists through the process of technology development, the production of knowledge and innovations, and the outcomes arising from their work.

Will technologists rise to the occasion to counter this alienation that is ultimately arising from the ideological hegemony of technology optimism and capitalism itself? Will they contribute towards changing the values that constitute capitalism, or replacing it with a different system? I believe so, and although I do not have any validation of this belief, I do trust that most technologists would not want to work where their labour leads to harmful outcomes. They would not want to lose their humanism, and would rather want to apply themselves towards genuine social good. Even the creative opportunities that white-collar technologists enjoy in their work and which keeps them motivated, would lose their charm once the misleading ideological veil of naïve technology optimism and the benefits of capitalism is removed and the reality is laid bare of how this is a carefully orchestrated illusion created through the apparatuses of media, educational institutions, and organizational settings, that are coordinated by corporations and governments. Technology should be used only for social good, and I discuss next how technologists can ensure that their labour is not misused and a new utopia of work can be created as doing social good. I hope that in these contemporary times, when technology paradigms can swing either way towards genuine social good on the one hand or towards control and exploitation on the other, technologists will see their work not as a job they have to perform unquestioningly or a profession they need to follow through rule-bound procedures, but as a calling that requires of them to speak up and act for their own sake and that of others (Jose, 2020).

Chapter 8

Overcoming Paradigms That Disempower

I have discussed so far that doing social good requires a transformation of unequal power relationships in society. Technologies intermediate many such relationships and need to be aimed at bringing power-based equality between its direct and indirect users. The role for technologists in designing and managing such technologies therefore treads into the political arena, but constraining organizational and economic structures, and misleading ideologies perpetuated within these structures, can prevent technologists from taking this political turn. However, not being able to influence the outcomes arising from their labour in creating positive social relationships, is anti-humanist and alienating for technologists themselves, and they will realize this sooner or later. Laws, regulation, and relying on consumers to make good choices, is not sufficient in itself. Technologists need to take proactive steps themselves, without which technologies will continue to be developed within the existing paradigms. In this chapter, I discuss some steps that technologists can take, and promising social movements, to do social good.

8.1 Changing the Status Quo

First, as would be obvious by now, technologists should realize that doing social good is not straightforward. Preventing unintended harms from technology is a challenging endeavour in itself, which requires an ethics-based alignment between different design components, and also going beyond design to ensure that management practices are built on the same ethics-based foundations. Overcoming the simplistic technology optimism mindset, or the limited rationalist approach to controlling technology, and instead embracing the romantic view of technology is essential for technologists to realize the importance of continuous steering of technology (Dahlbom & Mathiassen, 2016). Further, doing social good requires clarity on the terminal values and end-goals, and to arrive at a consensus on worthwhile values for society to follow. Technologists therefore need to expand their understanding of social problems and structural factors that cause these problems, including about unequal power relationships, to be able to work effectively to counter them. Building this understanding themselves, improved through discussions with

Technology and (Dis)Empowerment: A Call to Technologists, 145–159

doi:10.1108/978-1-80382-393-520221008

others, are important first steps before they can change the status quo (Dahlbom & Mathiassen, 1997).

Second, technologists should stop seeing themselves as individuals working towards isolated tasks laid down by their organizations, no matter how creative or exciting or challenging these tasks might be. Technologists should situate themselves as universal individuals and acknowledge the wider implications of their work in society (Foucault, 1984). Irrespective of where they are located in the technology production value chains, technologists should understand how their contributions relate to creating use-value in society through different technology projects where the output of their labour is deployed, and use this to evaluate whether their work is indeed aligned with social good. As discussed in the previous chapter, this humanism arising from being able to contribute towards positive social relationships in society shaped through the outcomes of their labour, is essential to avoid alienation. Only by understanding their universal situatedness will technologists be able to take effective steps to ensure that they themselves, the organizations where they work, and the governments of their countries and the world, target their energies towards social good.

Third, to understand the wider implications of their work, and to design or manage their technologies well, a familiarity with its users is essential (Suchman, 2002). As Spivak has discussed, this cannot be achieved by simply reading or learning about the users, most of whom occupy worlds very different from that of the technologists (Spivak, 1988). Not only are the complex histories and social structures of diverse users difficult for others to understand, with many underlying power relationships remaining hidden and invisible, a formidable barrier also arises from the typical exploitative role that most technologists have themselves historically played. Technologists often come from privileged sections of society, who, through the existing structures of the state and markets, have dominated over the subaltern, likely including those who are today the users of their technologies. This position of privilege has historically insulated them from having to listen to others or acquire the tools to understand them, and is likely to have contributed to an apathy of the non-poor towards the poor (M. J. Sandel, 2020). This is true not just in the United States or countries of the Global North. Gaps between the rich and poor have increased dramatically within countries of the Global South too, and technologists, who tend to be more privileged than the people who use their technologies, have little understanding of the contexts of their users (Piketty, 2014).

Participatory methods centred in collaborative relationships between technologists and users are likely to be more successful in equipping technologists with the necessary understanding and tools to positively affect society. Relevant participatory platforms or channels will be needed for technologists to gain such familiarity and to work alongside their users to involve them in design and management so that the technologies do not disempower the users and are able to lead to power-based equality. This will mark a transformation of the technologists to what Gramsci calls *organic intellectuals*, who understand the users, their society, and their social problems. Or what Freire calls *humanists*, who engage in the struggle of the oppressed through trust and respect for them as humans

(Freire, 1970; Gramsci, 1971). Only with this understanding will the technologists be able to formulate technology design and management strategies in participation with the people to solve important social problems.

Critical to this will also be for technologists to understand the political economy of technology which determines what technologies are adopted and for whom, so that technologists can adopt appropriate designs and forge suitable collaborations to address the social problems without causing further disempowerment of the weak (Seth, 2020a). As I have shown, most social problems have power-based inequality as their basis. Technologists will no longer be able to stay apolitical once they take sides to transform these power relationships. Technologists will thus transform into activists, and which will be a necessity if they are to embrace social good as a goal and to challenge the false consciousness imposed upon them (Dahlbom & Mathiassen, 2016).

Fourth, technologists should not expect that laws and regulations will easily fall into place to impose ethics-based foundations of social good as a goal for companies and governments. These entities have strong incentives to avoid or reduce the effectiveness of technology regulation, especially when their mutual interests coincide, as during the rise of surveillance capitalism (Zuboff, 2018). Technologists should, rather, try to bring in appropriate regulation in collaboration with society by understanding their problems, although this battle is hard since laws are shaped by those in power and used by the powerful to hold on to their privileges (Sawhney, 2021). Regulatory procedures further need to be responsive and agile to emerging feedback, and be empathetic in listening to feedback, especially from the less-powerful, but the effectiveness of these feedback loops needs considerable improvement to be able to rely on regulations alone for course correction (Seth, 2020b). Technologists will therefore need to develop their own ethos in the interim to guide themselves and their colleagues towards ethical actions. The unpredictability in technology projects, requiring careful management, will need ethics-based decisions at each step and can be guided only through a consistent ethos embraced universally and unambiguously (Frauenberger et al., 2017). Foucault gives the example of ancient Greeks who, in fact, saw ethics as an aesthetic element of existence, beyond what can be captured by law (Foucault, 1984). Emergence of a consistent ethos will however take time and it is likely to evolve as new knowledge and learning is acquired. Through a method of workers' inquiry, networks of technologists that facilitate archival and sharing of knowledge and strategies will help them find support structures to collectivize and build a common ethos (Huff & Rogerson, 2005).

Fifth, collectivism among technologists will be essential to enable them to discover and follow such a common ethos in their work: An ethos that acknowledges the limitations of technology optimism, conceptualizes and builds technology in collaboration with the users, is sensitive to the context of the users, and designs and governs technologies aimed towards genuine social good. The collectivism will give *structural power* to the technologists to challenge their organizations to embrace ethics-based foundations in its work, and to call out false claims. It will also increase the chances for responsible technologists to be able to influence the adoption of appropriate regulations by governments (Berg, 2008; Russell Einstein Manifesto, 1955).

The social nature of their profession to build technologies through cooperation, and the social networking platforms that they have built, provide technologists with both the skills and the tools to collectivize and coordinate with one another. This was even predicted by Marx as an outcome of capitalism's tendency to centralize the means of production through increasingly complex technologies. This requires building a more educated and skilled workforce that is capable of coordination and discipline, and such a workforce will therefore have better means to stand up to challenge capitalism (Adler, 1990). To have workers gain greater power within organizations, there is already precedence for mechanisms such as co-determination in Germany where employees have board representation in companies to influence decisions (Fox, 2018). With increased technology dependency in organizations and governments, white-collar technologists are indeed in a structurally powerful position to gain *institutional power* through collectivism and push for similar laws that can give technologists a greater say to influence the governance of their organizations towards stronger ethics-based foundations. Elizabeth Anderson in her book *Private Government: How Employers Rule Our Lives (And Why We Don't Talk About It)* draws crucial attention to the under-developed topic of governance within workplaces and how they tend to be similar to dictatorships (Anderson, 2017). Her main argument of employers to be more accountable to workers, as democratic governments are to citizens, is to ensure better working conditions and rights, especially for lower-wage workers. The insight she contributes is however more wide-ranging and valuable: that workplace governance should not be driven just by arguments of economic efficiency, as articulated in the *theory of the firm*, to have hierarchically organized systems of production to build complex products. Workplaces, rather, are often governed arbitrarily through misguided ideologies and even unaccountably on the front of worker rights, and therefore need better governance mechanisms that are inspired through a political understanding to institutionalize worker voices in the governance processes.

As the last several decades of trade unionism have shown, collectivism among workers will also need to extend globally across organizational and national boundaries so that the combination of globalization and neoliberalism is not able to create and exploit country-level differences in standards of doing social good (Hensman, 2011). Several tactics such as relocating production to different geographies, or introducing layers of firms through contractual or ancillary or subsidiary relationships, have been used in many industries to evade responsibility and accountability. Especially to ensure good labour standards, less success has been seen through self-regulatory methods such as conducting audits of working conditions in the supply chains of producers. Similarly, global regulation of labour standards has been hard to bring about. Even though the WTO regulates trade relationships for safeguarding capital, it does not mandate any minimum standards for labour. Labour standards have been left to the ILO, which, however, does not have any enforcement powers. As a result, neoliberalism within countries and the pressures of globalized competition between countries, has effectively led to a race to the bottom on working conditions especially for low-wage workers. Greater success has been seen through collectivism by global unions in transnational companies, and knowledge sharing and exchange of ideas between labour movements in different countries. This is quite possibly the case in the technology arena as well where regulations in countries of the Global South tend to be weaker than those emerging in the Global North countries, and highlights the need for global collectives of technologists to keep a check on their organizations and governments. Unions at global companies such as Amazon and Google now extend across national borders and indeed draw their power from this large reach.

Sixth, this collectivism should not be without its own ethos of democracy and commitment among the technologists, to avoid the pitfalls that trade unionism has otherwise seen. Collectives within organizations, or at the global scale, will need to democratically debate what values are essential to design and manage technologies for social good. A universal situatedness of technologists to be able to relate to different user groups will be required for these debates to be well informed. Additionally, technologists should resist the growing trend towards contractualization and commodification of their labour, or differences between permanent and temporary workers, since this will atomize them and reduce their associational power to bring changes (Ness, 2016). They should commit themselves to changing their organizations from within, and reduce job-hopping so that they can bring about these changes effectively. Finally, the interests and ideologies of white-collar workers are known to be somewhat incongruous with those of blue-collar workers, with the former being less favourable of redistribution, more individualistic than collective in their approach to rights and demands, and also more prone to an alignment with right-wing ideologies (Kocka, 1985; Owen et al., 1989). Ideologies that accept inequality and domination are not compatible with social good, and technologists should therefore find their political bearings on such topics which sometimes polarize them (Arndt, 2018). Technologists will need to overcome these differences to build a universal alignment for genuine social good. It is heartening to see white-collar unions now emerging in the technology industry that see their role as not only defending working conditions for technology workers, including for blue-collar employees in the company, but also as ensuring a responsible use of technology (Alphabet Workers Union, 2021; Conger, 2021).

Seventh, technologists can use contractual rights to define purpose restrictions in the use of intellectual property (IP) that they create, whether as individual creators or during their employment. Both copyrights and patents are considered as an output of individual thought, and require specific employment contracts for the rights to be assigned from the original creators to their employing organizations (A. Moore & Himma, 2018; Syreng & Boyd, 2020). Most employment or retainer contracts contain standard clauses that specify complete assignment to the organization of any IP created by the individual during the term of employment or the project. Individuals working within organizations, therefore, have the prerogative to negotiate on their employment clauses to define purpose restrictions such as to use technology exclusively towards social good. Such negotiations can be done by technologists situated anywhere in the technology production value chain, to enforce that the technology artefacts on which they have worked are used only in technology projects that contribute towards social good. Even novel ethical source software licenses have emerged on similar lines, such as the anti-capitalist license, which allows use of the software only in organizational setups where wage labour is not used, the non-violent license which disallows use of the software for any form of violence or environmental harm, and the Hippocratic license which follows the UN Human Rights principles (Ethical Source Licenses, 2020). While enforceability of these licenses can be difficult, they can serve as an important signalling mechanism for technologists to clarify their ethics. Further, transparency technologies such as blockchains could possibly be used to track software use across value chains, and seek self-certifying endorsements from software users on their use of the software towards ethical goals.

Eighth, a wider societal partnership will be essential to counter the dominant capitalist ideology of self-interest and competition with that of cooperation, of consumerism with that of environmental consciousness and to meet genuine use-values, of high modernism and technology optimism with that of responsible deployment of technology, and to identify key values for social good. Such a societal partnership will need technologists to reach out to society to improve its awareness about technology risks, and provide access for society to global democratic communication platforms on which the values that shape the design and management of technologies can be freely and openly deliberated. As discussed in Chapter 6, communication starting with discourse within small groups can grow to a societal scale and alter the ideological hegemony in which the world today is trapped (Dijk, 1989). Technologists already have a rich experience of building and managing large information and communication platforms such as Wikipedia and Github, they are aware of what can go wrong, and how to impose appropriate design elements and management processes to ensure responsible use. Such platforms for meta-social good that can ensure power-based equality in their functioning, and aim to build an understanding and respect for diversity and plurality, can form the foundation for technologists to collectivize with one another and with the wider network of users of their technologies, and cooperatively work towards identifying values to define social good. This aspect is discussed in detail in the next chapter.

Ninth, following through with ethics-based foundations to do social good will not be straightforward. Entrepreneurs from among the technologists, or organizations and governments that technologists are able to influence, will need to compete with their counterparts who may not be aligned with genuine social good goals. Mobile Vaani, for example, faces tough competition from news and information providers who do not espouse similar values of inclusive access and pluralism, and hence finds it hard to compete with lower-cost initiatives that pursue simpler goals (Seth, 2020d). Platform cooperatives for ride sharing that emerged to counter the individualistic competition and self-interest of capitalism by demonstrating alternate collective structures, have to compete with behemoths such as Uber that can easily wage a price war to drive out small competitors. Distributed applications for social networking, like Diaspora and Omlet, that emerged to counter centralized platforms like Facebook, seem to not be able to keep up with the performance and user-friendly conveniences that Facebook is able to build more easily because of its much larger resources (Lam, 2014). The Lucas

Plan of 1976 created by collectivised workers to transform their company towards using its capabilities for humanitarian goals (see box on Architect or Bee), was ultimately rejected by the management who had the final say in what products were produced (Cooley & O'Grady, 2016). Sustaining such collective efforts themselves may not be easy when groups grow and free-riding effects begin to dominate (Olson, 1965). As is evident from these examples, it is not easy to do social good. A combination of grit, strategy, consumer awareness, and new forms of financing, are needed if social good projects are to compete with other projects that provide similar functionality but without the same ethics-based foundation.

Tenth, at heart, the key change is to transform the current dominant paradigm of how technologies are conceptualized, designed, built, managed, and scaled. Overcoming this hegemony will not happen overnight and requires revolutionary movements on many fronts: inside organizations, at a societal level through democratic means to change how people perceive technology, and by educational institutions to build more aware citizens and technologists who want to work for social good.

Although clear answers may not be evident, many such revolutions are already happening and breadcrumbs do exist that could pave the path for the proliferation of social good projects.

Free and Opensource Software (FOSS) movements: The growth of open-source systems, for example, has demonstrated that rapid innovation can emerge through shared knowledge and resources without any expectations of material gain. Such materialistic incentives of wealth accumulation secured through private ownership of the IP or assets are not the only incentives that motivate people to innovate. Rather, incentives of solidarity to work in cooperation with others, or the imperative for social good as a key characteristic of humanism, are strong motivations in themselves. Another aspect strongly demonstrated by the FOSS movements is the relevance of transparency: systems that are open for examination and auditing are considered more trustworthy by people than closed systems that claim correctness in their operations (Lessig, 1999). This view is increasingly important to keep a check on AI and algorithmically driven systems for which arguments are often made to justify closed systems by citing security concerns and proprietary trade-secrets.

Venture communism: This has been suggested as a new form of financing for cooperatives, to enable them to compete with large corporations (Kleiner, 2010). This is intended as a first step to counter capitalism, by building an exclusive commons space that acquires any necessary material means of production from the external capitalist world, but utilizes them internally in a cooperative manner to produce goods that are needed by other members of the commons. The only way for people to gain membership is by contributing their own labour. As this commons grows and builds its own economy centred on labour, it may not need material assets from the external world any more, and can in fact start producing goods for the rest of the world to bring more and more of it into its organizational fold. New software licenses such as the Peer Production License have been proposed on these lines, to strengthen the commons so that it can support ethical economic entities such as cooperatives (Bauwens & Kostakis, 2014). Similarly, concepts of data trusts and data cooperatives have been proposed to safeguard against data misuse by establishing fiduciary entities which can license the data to external entities for specific purposes that are acceptable to the members based on governing principles of the trust or cooperative (Ada Lovelace Institute and AI Council, 2021). The contemporary focus of capitalism on immaterial factors of production such as code, data, and knowledge, which gain productivity when they are freely shared, instead of being constrained to silos within corporations, may indeed lead to the commons as becoming the more accepted form of ownership of this immaterial property, and introduce a new contradiction to capitalism (M. Hardt, 2010). Such an approach should be favoured instead of proposals to convert data into a tradeable commodity (R. Pal et al., 2021). Proposals like these tend to attach a price to privacy itself and give markets the freedom to seek out the lowest price, which invariably only further dispossesses the poor.

Commons-based management: Ostrom showed that communities can indeed evolve methods to collectivize and manage their commons responsibly, without having to involve external experts or governments (Ostrom, 1990). She identified several principles that would increase the likelihood of self-organization, enhance the capabilities of the community members to sustain the efforts over time, and solve problems to avoid the tragedy of the commons. Mechanisms that improve observability are able to strengthen *associational power* that otherwise can be weakened in large groups. Graduated sanctions against rule-breaking, and observability of the sanctions, improve accountability and discipline among the members. Democratically built context-sensitive rules for sanctions, the distribution of resources, fair handling of exceptions, and fast and fair conflict resolution, improve trust of the members in the operation of the collective. Transparency in allocations, bookkeeping and regular measurements of usage, and local knowledge sharing structures, further improve the commitment of the community towards the rules. Polycentric or federated governance structures built on these principles have been able to scale to large populations, where independent groups use the same setup to manage relationships between themselves as they use to manage relationships between individuals within a group. Drawing from these principles, similar collective structures can facilitate technologists to both – network with one another to build technologies for social good and work with the wider society to evolve rules for usage and governance that avoid harm and do social good.

Appropriate technology: Schumacher drew attention to appropriate or intermediate technology as that which can be built and managed by local populations themselves (Schumacher, 1973). This reduces their dependency on large technology providers, and builds their own skills and institutional structures to manage the technology. This philosophy finds resonance with participatory design methods which place emphasis on capacity building of the participants so that they can participate as equals in the technology design and management process, further layered with terminal values of social good to prevent oppression and disempowerment of the weak. This is often seen to be in conflict with the platform approach that is increasingly popular in current times. The platform approach imposes its own logic on the users to constrain their actions along affordances determined by the platform providers, rather than provide users with the tools through which they can steer the use of the platform according to their needs. Further, it makes the users subservient to the platform and the platform providers. The design principles of appropriate technology argue just the opposite, being aimed towards putting control in the hands of the local community to determine the means and ends addressed by technology, and are therefore more in agreement with the ethics-based foundation for social good as bringing power-based equality. Even if technologies are built externally, Schumacher's arguments remain valid in having local communities control these technologies, because only then will users begin to understand its capabilities and limitations, and thereby strengthen local institutions to design management processes to control it.

Role of civil society: Civil society groups have succeeded in drawing attention to the problems of rising inequality and global warming, and voices are rising to introduce new values into capitalism to reduce inequality and emphasize care for the environment. How successful can this be in changing capitalism is yet to be seen. It, however, cannot be denied that civil society is indeed becoming a strong voice to keep capitalism in check. An alert civil society is creating a stronger consumer demand for ethics-based foundations in technology systems by raising consumer awareness about the harms of technology on the wider society (Irwin, 2015; K. White et al., 2019). Similarly, labour movements, social movements for gender equality and non-discrimination, and growing demands for civil liberties, have come together to uphold democracy and keep a check on capital. This was demonstrated with the eventual repeal of unjust farm laws in India which had been earlier rammed through parliament by taking advantage of the COVID-19 pandemic to avoid debate, and similar movements have registered their dissent against the lack of adequate safeguards for farmers in agricultural technology proposals such as the Agri-stack (T. Ali, 2021).

Information ethics: Scientists and innovators are recognizing the dual-use nature of information technology (IT), especially with new forms of harm such as in political applications of surveillance and neuroscience technologies to manipulate the beliefs and behaviour of people (Harris, 2016; Mahfoud et al., 2018). They have called for stronger regulation of IT. This may lead to preventive forms of regulation for IT, as is the norm in bio-ethics, where products are rigorously tested before they can be released for the public. Alternatively, highly managed forms of regulation may be recommended such as for nuclear devices, where detailed instrumentation and regimented rules are laid down to avoid disasters. Floridi's conceptualization of humans as information beings who are shaped by information, and thus need to be protected from misinformation, leads to the same conclusions: to carefully evaluate technologies that mediate the flow of information and that can harm the informational constitution of human beings (Floridi, 1999).

In short, none of the social structures can be taken as given. Technologists should discover new structures to carve out a path as moral exemplars, through resourcefulness and dedication, aided through collectivism (Huff & Rogerson, 2005). A clarity of purpose, the humility to continuously correct course to steer their innovations, a commitment to social good, and gaining political power through collective means to exercise their judgement, can help technologists counter the dominant paradigms that produce disempowering technologies. Technologists should aim to architect a new system, and not remain like a bee that works hard meticulously and with dedication but within the specifications laid down by existing systems, thereby preventing a departure from the status quo (Cooley & O'Grady, 2016).

Architect or Bee? The Need for Revolutionary Movements

> A spider conducts operations that resemble those of a weaver, and a bee puts to shame many an architect in the construction of her cells. But what distinguishes the worst architect from the best of bees is this, that the architect raises his structure in imagination before he erects it in reality. (Karl Marx, Capital, Volume 1)

Facing competition and the need to restructure, an automotive and aerospace parts manufacturer in the UK, Lucas Aerospace Corporation, announced in 1976 a plan to retrench thousands of workers (Cooley & O'Grady, 2016). It was then that under the leadership of Mike Cooley, the workers came together and put up a plan called the Lucas Plan, about using the capabilities of the company and the skills of the workers to produce socially useful products. The plan eventually was not accepted by the management, but this was a prominent event that demonstrated how workers could come together to re-architect an economic and technological system that was not solving meaningful problems for society, and to produce useful employment at the same time. The design of all proposed products was done by the workers themselves by following human-centred design principles, to show that workers could re-possess the ground for decision-making that had become restricted only to planners and managers. The design and development process also brought intellectual satisfaction to the workers, which machines otherwise tend to diminish through their de-skilling characteristics (Braverman, 1974). Most importantly, the Lucas Plan showed that workers were not content to restrict their role as simply wage seekers engaged in production, but that they could impose political and ideological choices on their companies to shape new kinds of social relationships, which they found to be more relevant and socially useful rather than purely commercial or consumerist relationships. Methods such as co-determinination can give workers more power to change their organizations from within and ensure that any future Lucas Plans are actually adopted and implemented (Fox, 2018).

The Lucas Plan is not a solitary beacon. Modern movements like Fablabs, hackerspaces, open-hardware designs such as RISC-V and Raspberry Pi, peer production, etc. emerge from similar ideologies of having people explore new socially relevant problems and design solutions collaboratively (A. Smith, 2014). Unlike the Lucas Plan, however, such movements are community based and exist outside organizations. A similar community-based approach has also been taken by the older Free Software and Opensource Software movements which aim to force corporations to provide opensource software that users can freely change and share, as well as be more open in their production process. These movements aim to counter the monopoly of proprietary software through alternatives created in a free, open, and community-driven manner. Although the movements have had many successes, struggles persist as is obvious from inherent ideological differences within the movements themselves between the values of being *free / libre* versus *open* (Liu, 2018; Stallman, 2007). Many opensource components that provide frameworks and commodity tools are used by companies like Facebook and Google to build platforms in which, however, the algorithms and design processes remain opaque and non-free. Building alternative platforms that are more free is also difficult for many reasons, ranging from capital requirements to network effects that make it hard to displace existing proprietary platforms. The Free Software movement is therefore considered as having lost out to the Opensource Software movement. The Opensource Software movement, on the other hand, is considered by many as having degenerated into providing free labour by volunteers which ultimately benefits corporations (B. J. Birkinbine, 2020). Carefully crafted strategies have indeed emerged within the Opensource Software movement for co-existence of commercialization and volunteer work, but they do not seem to be stable enough and require constant tweaking under changing circumstances and technology landscapes.

New movements arising from the commons philosophy aim to draw a clearer line between the commons and the market, and to grow the commons by drawing value from the market into the commons (B. Birkinbine, 2018). It suggests that capital's strategy of accummulation by dispossession can be reversed through a strategy of *commons pooling by capital dispossession*, by identifying value circuits of production within the commons, and between the commons and the capitalist market, which can be managed so that the commons become self-reliant and have more to offer to the market than having to take from it. Several strategies are being developed on these lines: pooling of resources between different commons such as the land commons for food and digital commons for software, value accounting tools using blockchains to make commons pooling easier, and bringing along the state as a partner to support the commons instead of capital (Bauwens & Niaros, 2017). The creation of democratic institutions to manage the commons and avoid tyranny in its administration is clearly essential for such movements to succeed. This approach has attracted criticism too. The argument is that values such as transparency and openness are sufficient to reform capitalism from within, and that adopting capitalism's strategy of accumulation to defeat capitalism on its own terms is ultimately anti-humanist because such a strategy will end up exploiting workers who may be stuck in the market economy not out of choice but because they do not possess the resources to pull out easily from regular wage-seeking labour (Dafermos, 2016; Papadimitropoulos, 2017). A clear solution may not be obvious, but commons-based movements are probably the most active spaces today searching for new directions.

Another prominent counter-movement against exploitative use of technology is that of the cypherpunks who aim to use cryptography as a tool to protect individual autonomy that is threatened from surveillance by the state or corporations (Rogaway, 2015). Solutions like PGP and ToR are meant to give people the ability to conceal their communication messages and identities. WikiLeaks is meant to disrupt entrenched power that is maintained unlawfully or through hidden practices that society is likely to reject once it becomes aware about these practices. Snowden's revelations are also consistent with the cypherpunk movement, and demonstrated the effectiveness of this pathway of revealing problems to society and then letting democracy do its work of rejecting values that it considers undesirable.

Revolutions are therefore needed on both fronts to architect a new paradigm for technology: through collective means inside corporations, such as what the Lucas Plan attempted to do, and through community-based means outside corporations, as what the Free Software, Opensource Software, Digital Commons, and cypherpunk movements aim to do. Technologists clearly have their work cut out, to experiment and learn to build new paradigms for their work.

8.2 The Future of Technology for Social Good

The world today faces grave social problems, ranging from inequality, to poverty, exploitation of the poor, global warming, access to social protection, and threats to pluralism and diversity, among others. I have shown that power-based inequalities are behind many of these problems, and collective structures are necessary to counter these inequalities. The dominant capitalist ideology of individual self-interest, competition, and denial of unjustified inequalities manages to suppress collective action, but technology can be a fundamental part of the solution if it is suitably deployed for social good, including to challenge the dominant capitalist ideology by introducing new values within it or replacing it with a different system. However, simple populist and orchestrated ideologies of technology optimism instead lead to mis-deployment of technology, which exacerbates social problems even further. I have outlined a tall order for technologists to understand these complexities, and to counter the ideological hegemony of technology optimism so that technology can be deployed more appropriately and with a focus on social good. I have further shown that ethics-based foundations are necessary to reason about social good.

Similar calls for an ethical examination of technologies, and to use technologies for social good, have been made by others too. Shneiderman introduced the concept of Social Impact Statements (SIS) that should accompany any technology development, and specify the user communities for whom the technology is intended, establish training requirements for use of the technology, specify potential negative side-effects to guard against, and institutionalize monitoring procedures or usage and outcomes throughout the technology's lifetime (Shneiderman, 1990; Shneiderman & Rose, 1996). Unwin draws attention to how the naive belief in technology optimism is shaped by the capitalist elite who profit from continuous technology innovation irrespective of its ethics, and states that technology should only be created for the poor, otherwise inequality will only get worse (T. Unwin, 2018a, 2018b). Ethics needs to underpin any technology and Rogerson places faith in the millennial generation to take ethics more seriously (Rogerson, 2015). Moore and Hodson draw emphasis to expanding academic courses on technology ethics with content on contemporary political and social problems so that technologists can target their innovations more precisely (Hodson, 2010; J. Moore, 2020). Green argues that technologists who claim to be apolitical are in fact taking a conservative political stance themselves, that aligns with the status quo of the prevailing systems (Green, 2020). Rogaway discusses that the ethic of responsibility among scientists that emerged prominently after the development of the nuclear bomb, has declined in the field of computer science during contemporary times due to the prevalent ideology of technology optimism, apolitical academic research, and distributed accountability arising from the creation of differentiated roles in technology design and management (Rogaway, 2015). Even within the highly mathematical field of cryptography, he shows how centralization in the design of a technological artefact, or in the power relationships between different stakeholders shaped by the technology through its deployment in the real world, is ultimately a political decision that needs to be shaped by the values of the scientists developing the technologies. Balabanian considers engineers as driven by professional principles rather than political values, and suggests that whisteblowing by engineers to call out problems in technological systems that can impact public health and safety can help raise regulatory attention to steer markets (Balabanian, 2000). I have tried to push these arguments further by highlighting that technologists are extremely well placed to bring about a transformation in themselves, their organizations, and their governments, so that technology is used only for social good, and I have outlined several steps that are essential to doing this successfully. Below are a few questions that can help technologists introspect the outcomes arising from their work.

- How does your work ultimately enable different applications, services, and products that people use?
- Is there a way for you to interact with these users? Or are there other ways for you to understand how users are affected by the products resulting from your innovations?
- Are the outcomes positive for some people and negative for others?
- If you do not like some of these negative outcomes, does it bother you that your work is leading to outcomes that you do not agree with?
- Is it possible for you to change anything in your work to ensure that more desirable outcomes arise? Or for your team to change its ways and membership so that it can pay better attention to these negative fallouts? Or for your organization to adopt new practices internally, or advocate to customers and vendors upstream or downstream in its production value chain to bring changes at their end?
- Are there others in your organization who agree with your viewpoints?
- Do you and like-minded colleagues have pathways available in your organization to put across your viewpoints?
- Would these be heard and acted upon by the management?
- If they would be opposed or dismissed, explicitly or through inaction, do you understand the broader political economy to be able to reason why?
- Could the organization's management have brought about the changes but did not, or were their hands tied because of wider economic and governance systems in the world that would impact competitiveness or finances?
- If appropriate action was feasible but still not taken, can you think of ways to have more power in organizational governance and decision-making to ensure that meaningful action is taken in the future? Can it be done through enforcement of IP rights to define purpose specifications for which the output of your labour can be used? Can participation in organizational governance be gained through collectivization with your colleagues?
- If appropriate action was not feasible because of imperatives imposed by the wider systems of the state and markets, would you like these systems to run differently?

- Do you think you can affect such change in a democracy? Are there social movements in your country or around the world that you are aware of, which are pushing for similar changes?
- Would you lend your support to these movements by collaborating with them to propose new regulatory structures and laws, or by building alternate systems and ecosystems, or by monitoring existing systems for greater accountability?
- Do you think there is recognition in the wider society of the need for these changes, so that there is hope in democratic setups that laws and regulations can arise for organizations to take more responsibility or for governments to create better policies?
- If not, do you think you can change that by building greater engagement with users to make them more aware, or by collaborating with them in campaigns for change where you can make strong contributions because of your deep technical knowledge about alternative design elements and management practices that can be built?
- Or do you feel that it is not your job to engage in campaigns? If so, what would you suggest should be done to ensure that desirable outcomes arise from technology?

My hope is that many technologists will feel that it is their job to bring change. The future I envision, if technologists are indeed successful in efforts to reshape technology paradigms, is a world of technologies that can be controlled by their users, with an opensource and transparent design, cooperatively owned or marketed through radically altered capitalistic values that are more inclusive, not designed and managed by large companies but by a digital commons or democratic coalition of local institutions of people, who have the required skillsets and knowledge to build and operate the technologies, are trained in technology ethics, they take responsibility to steer the technologies away from harmful outcomes, by following participatory practices of design and management, with constructive debates to define specific social good goals, and to use technology only for social good that can bring power-based equality in the world. Such a future will imply that all stakeholders – technologists, citizens, and people – are aligned in their values at a societal level and committed towards social good. A networking infrastructure can provide the foundation for communication among the stakeholders, to debate and deliberate, listen to one another, respect plurality, understand and empathize with others, build new knowledge for suitable action, and offer opportunities for collective action to counter power inequalities whenever they emerge. Technologists should aim for such a future.

In the next chapter, I describe what such a communication infrastructure could look like, to enable society to discover values for social good through which they can impose their control over technologies, and for technologists to build and manage technologies in accordance with these values.

Chapter 9

Societal Participation

So far, I have discussed that for technologists to do social good and re-acquire their humanism, or to work towards meta-social good projects to transform entrenched power structures in society, requires participation by society to determine the values that should define social good. Such participation requires a networking and communication infrastructure for deliberation that respects diversity and plurality, includes marginalized groups, and has situated itself in a powerful position to affect institutional change. The values for social good identified and espoused by society will then be able to make their way to markets to alter capitalism with a new ethic, or create corresponding regulations, or bring about new structures altogether for the functioning of the state and economy. Communication infrastructures are needed also for technologists to gain visibility into the problems faced by users who are directly or indirectly affected by their technologies, as well as to inform society about technological risks and help collaboratively to build regulatory methods. In this chapter, I discuss essential features for such communication platforms, the shortcomings of many platforms available today, and how commercial mass media has been co-opted by the state and capital rendering it ineffective to both represent society as well as to affect institutional change. Such a meta-social good project of communication platforms to facilitate values discovery and their institutional adoption is essential for the success of any social good project. It can also serve as an essential infrastructure for democracy itself, through an embrace of pluralism and power-based equality for participation. Technologists can contribute towards building such platforms that promote plurality and democracy, which in turn will empower society to impose effective social control of technology so that it does not disempower the weak, and additionally help technologists to not get alienated from their own labour.

9.1 Media and Deliberation

9.1.1 The Need for Participatory Communication

The discovery of values for social good in a consultative manner is a function of democracy, to determine the underlying ethics with which society wants to govern

Technology and (Dis)Empowerment: A Call to Technologists, 161–181

Published under exclusive license by Emerald Publishing Limited
doi:10.1108/978-1-80382-393-520221009

itself. Theories related to deliberative democracy provide a useful starting point to understand this discovery process and highlight the relevance of participatory communication platforms that technologists can help build.

Carlos Santiago Nino in *The Constitution of Deliberative Democracy* argues that only morality can be the basis of law (Menendez, 2000). He suggests that morality is discovered as a consensus of values or rights arrived through a social practice and meant to guide towards conflict resolution and societal coordination to achieve complex social goals. Such a social practice requires certain principles for discussion: impartiality, rationality, and full knowledge of relevant facts. Nino states that deliberative democracy is the only legitimate procedure of decision-making for this purpose. This is because of the capacity of deliberative democracy to expose citizens to the different opinions and interests of others, even leading to a transformation of their own preferences in the process. He gives the example of how values of social rights such as the right to health, to proper shelter, and to a fair income, that are now considered as standard practices in a good society, actually emerged over many years through a consensus that formed with improved awareness, reasoned discussion, self-reflection, and impartiality. These social rights are now enshrined in the constitutions of many countries and form the basis of their law. Law constantly undergoes reforms to reflect evolution in these values and rights. A crisis of democracy for Nino lies in the growing apathy among people and their lack of concern for public affairs, which reduces their participation in deliberative processes, and thereby impacts the effectiveness of democratic procedures for values discovery and societal learning which are essential for law making (Oquendo, 2002).

Habermas identifies a similar link from participatory communication – which produces understanding (termed as *communicative action*) – and the use of this understanding to arrive at a consensus on shared values, which, in turn, prompts ethical action (Habermas, 1996). Values are considered as insights realized through reason and reflection on experiences, and are open for discussion. Communicative action is distinguished from other forms of communication such as *open strategic action* where coercion or unequal power relations are used to influence the communication process, or *closed strategic action* when these biases may be hidden. Such strategic communication impacts the ability of people to gain a genuine understanding through discussion and thus compromises the deliberation process. The public sphere is the primary site that Habermas identifies where deliberation happens, and a crisis of democracy for Habermas lies in the transformation of the public sphere from a site for hosting ideal communicative action to a medium that has been colonized for strategic action by the state, corporations, or the bourgeoisie (Habermas, 1989).

Other fields also put a strong emphasis on participatory communication processes. Cybernetics sees communication as key to establish information flows carrying feedback signals in a wide ranging of domains from automated guidance of missiles to communication for governance (Duetsch, 1963; Wiener, 1950). A visionary project was initiated in Chile by Stafford Beer, recognizing the need for feedback from citizens to bring inclusion for participation in democratic processes, which would add *variety* in making governance decisions and help counter any ideological hegemony (Beer, 1975; Espejo, 2014). He modelled citizens as communicating with one another through media platforms to create variety and inform the governance process. Effective communication required the legitimacy, authenticity, and competency of participants, and decentralized systems which allowed the discovery of new variety were favoured over centralized systems which restricted variety. Noisy communication or broken feedback loops were considered as leading to an increase in the state of disorder of the system (Wiener, 1950). Cybernetics thus considers the sanctity of communication as critical for a system to achieve its goals.

A similar approach is taken by Floridi in information ethics, where, instead of placing the responsibility for actions on free agents as in deontological ethics, or taking a consequentialist patient-centric view as in bio-ethics, information itself is considered as an entity that requires protection (Floridi, 1999). Information is seen as having rights of its own to persist and flourish, and enrich its existence and essence. This is measured as the entropy of the aggregate infosphere, being composed of both mechanical as well as human agents, both of which are considered as information beings constituted of information and shaped by information. Error-free communication, data protection, and other information processes are evaluated based on whether or not they increase the entropy in the aggregate infosphere. Based on this principle, information ethics has been applied to provide guidance on privacy policies to adopt in different scenarios, highlight the negative impact of misinformation in the public sphere, the relevance of communicative action to improve understanding, and deliberation as a means to arrive at consensus on decisions impacting the infosphere.

Media, communication, and deliberation are therefore essential for societal learning, values discovery, and law making in a democracy. This in turn is vital to impose effective social control over technology and ensure that it does not lead to negative fallouts. Technology governance therefore needs democracy – for society to discover the values that are important to it and which technology projects should respect – and democracy in turn needs technologies for participatory communication to discover these values. Technologists should contribute towards building participatory communication platforms for democracy so that values for social good discovered through this deliberation can be applied to design and manage technology projects for social good.

9.1.2 Broken Mediums for Participatory Communication

Given the relevance of participatory societal communication for decision-making and democratic governance, and various deemed pre-conditions for its effectiveness, as envisioned in a deliberative democracy, or from a cybernetic standpoint, I next discuss what hampers current systems for societal communication. *The Structural Transformation of the Public Sphere* by Jürgen Habermas provides an account of the changes in Western Europe and the United States in the use of media by people to discuss and share information, and influence government (Habermas, 1989). Habermas views the public sphere as composed of multiple forms of media that serve as a means for rational critical debate through which society forms a public opinion. During the seventeenth and eighteenth centuries, such rational debate would happen in publicly accessible salons and coffee shops to build public opinion. With the coming of the French revolution, democracy was conceived as a vehicle through which this public opinion could be channelled into making laws, by electing representatives and having them carry the debate forward into parliament. However, several developments impacted these two functions of the public sphere, of forming public opinion through debate and then channelling it into laws. First, the individualization of society with growing capitalism and class structures made people closed to interacting with the rest of the public. This led to the creation of many publics, with apathy or conflict between them and the absence of any venues to bring them together for rational critical debate. Second, the rise of the written literate form of communication for governance processes excluded the non-literate from participation, leading to further fracturing of the public sphere. Third, with the development of the printing press, the venues of interactive debate were replaced with one-way communication platforms of the mass media. Capitalism appropriated this emergent vehicle of mass communication by turning it into a medium for advertising to drive consumerism. Fourth, the rise of the democratic state led to politicians competing with one another to use mass communication tools for influencing public opinion for their electoral prospects, rather than the other way around, that is, using communication technologies to listen to the people and represent them more effectively. Limited in their abilities to rein in capital and provide welfare services, governments also began to impose censorship and editorial control to curb inconvenient rational critical debate.

According to Habermas, the link, therefore, of politics as a vehicle to channel public opinion into law, and of genuine public opinion as arising only from an underlying morality, was broken. The public sphere degenerated into serving different private interests, of corporations or politicians or bureaucrats, relying on the publicity and spectacle creating abilities of these actors rather than underlying guiding values of morality or reason. The function of deliberation in the public sphere which was supposed to forge a social contract through consensus and by being grounded in morality, to then be shaped into law through the institutions of democracy, was lost to an economics of publicity to influence consumer behaviour. Society itself degenerated into a consumer of content than a producer of opinion – "a culture consuming rather than a culture debating public". Habermas termed the dominant majority opinions shaped in this manner as *social opinion*, to be distinct from genuine *public opinion*. Governments and corporations however routinely masquerade social opinion as public opinion to justify their actions, even though actually the required democratic function of the discovery of morality through rational critical debate to then shape law, is no longer effective.

A broken public sphere thus impacts democratic governance, including of technology. Further, it also affects the cultural capability of society to co-exist in a pluralist and mutually supportive manner. Habermas defined the *lifeworld* of people as their conception of cultural practices that are reproduced through the public sphere. A compromised public sphere that has been colonized by corporations and the state results in a colonization of the lifeworld as well, and leads to issues such as a growing apathy of the non-poor towards the poor, being complacent about the rights of others, growing consumerism, and environmental disregard. This weakening of social relationships of communication leads to alienation and loss of humanism, just as how Marx considered alienation as emerging from weakened social relationships constructed through capitalist production processes (T. Young, 1985). Fuchs, in fact, suggests that communication itself can be considered as a production process, and is shaped by the forces of capital that make this production process exploitative, coercive, dominated by the powerful, and ultimately alienating for humans (C. Fuchs, 2020). A compromised public sphere thus devalues democracy, pluralism, and representation of marginalized groups, and leads to their disempowerment.

Habermas further argues that degenerated public spheres in the capitalist world were also weakened with the growing rationalist orientation of society which raised questions of whether knowledge, morality, or reason to guide laws could even emerge from the public sphere. Public opinion was just opinion and not necessarily correct in a rationalist sense with its conception of searching for universal truths. Even new forms of rationalist governance were advocated, such as *open societies* which purposefully experimented and discovered these truths (Popper, 1945). All this ultimately weakened the relevance of public opinion itself, and the belief that new truths could emerge from the public. Rather than having a public sphere that allowed different forms of knowledge to grow and engage in debate with one another, the public sphere became a means to impose hegemonic structures for knowledge and suppress emancipatory knowledge to challenge the dominant ideologies (Foucault, 1984). At the same time, rationalist approaches that were successful in the natural sciences to understand the world and guide technology development, were, often found ill-suited to govern societies. This has been amply demonstrated by James Scott, Arturo Escobar, David Graeber, and others (Escobar, 1995; Graeber, 2015; Scott, 1998). The failure of the bureaucratic route of using rationalist reasoning in governance, as well as colonization of the public sphere that compromised its ability to foster genuine public opinion, reinforces the need to see democracy in terms of its true function of deliberation, agreement and agonism, listening to the weak, and to fall back on morality as the basis for law.

Habermas' analysis of how public spheres are transformed through capitalist ownership structures and rationalist approaches to governance, and their interplay with the lifeworld, finds echoes even in contemporary times where mass media in many countries is not able to provide a satisfactory medium that supports vibrant public spheres for deliberation. Growing concentration in the media industry globally has reduced media pluralism, which impacts the ability of the media to improve political accountability (Arsenault & M. Castells, 2008; Prat & Stromberg, 2013). Further, the orientation of commercial media to chase audience numbers, instead of improving the diversity of views presented to people, rather increases polarization by partitioning audiences into camps of mutual opposition (Mullainathan & Shleifer, 2005). Funding of the media by government advertisements, coercion by threats to the safety of journalists, and weak judiciaries that are not able to prevent censorship, also becomes a route for media capture by the state (Ninan, 2009; Schiffrin, 2017). A compromised media thus becomes ineffective in making voices of the marginalized heard by the wider society (Dreher, 2009), places more emphasis on sensational themes than social justice (Jansen et al., 2010), and uses agenda setting and framing techniques to further commercial interests or political alignments instead of the formation of genuine public opinion by the people (Leeper & Slothuus, 2018). The pre-conditions for deliberation are not met because citizens who may wish to learn about the views of others through the media are now not able to hear unadulterated expressions and viewpoints, or justify their own positions, through which they could have, by mutual respect of others, re-evaluated or revised their own preferences. Further, media control not only diminishes the ability of society for deliberative and democratic decision-making, its influence through entertainment and advertising also accentuates cleavages in society by catering differentially to target audiences along dimensions of religion, culture, and consumerism (Chakrabarti, 2014; McGuigan, 2015). These cleavages leave society more vulnerable to political manipulation, and thereby weaken democracy itself.

Given the limitations with current structures of the media, the rise of the Internet and digital participatory media platforms was initially seen as reinvigorating the public sphere and the lifeworld, and reducing the growing alienation in society (Salter, 2007). The richer nature of the medium, being more interactive and participatory, could improve user engagement, their ability to learn from one another, and spawn new uses to which such media could be deployed (McLuhan, 1964). It could counter the social opinion shaping power of the mass media. It could lead to genuine public opinion through deliberation and debate, and keep a check on power. It could neutralize the hegemonic and colonizing forces that weaken public spheres. It could lead to an information-rich society in which all forms of knowledge are produced and balanced against one another. The relevance of such a media revolution centred on interactive and democratic communication was envisioned even in the pre-Internet era, to

> enlarge praxis, promote community, maximize peace, advance social justice, respect the physical environment, instead of continuing to reproduce social inequality between rich and poor nations; rich and poor businesses; rich and poor classes; reproduce the privileged and unprivileged gender and ethnic groups; and the dangerous instabilities that these bring. (T. Young, 1985)

Realizing these merits of the Internet has, however, not been straightforward. The design of most social media platforms has created usage norms which, in the case of differences between users, leads to the silencing and disengagement of one side rather than a civil debate (Grevet et al., 2014), formation of echo chambers and heightened polarization rather than understanding (Conover et al., 2011), and filter bubbles that reinforce biases rather than expose users to different viewpoints (Bakshy et al., 2015). The politicians, governments, and corporations on which participatory media systems were supposed to keep a check, have ended up utilizing social media platforms for their own marketing and branding (S. Chakraborty et al., 2018). Gatekeeping in mass media to control the coverage of diverse views has been replaced by hierarchies on social media platforms where celebrities, influential users, and influence maximizing algorithms have become the new gatekeepers (Wihbey, 2014). One hierarchy has been replaced by another, and a divided public with little cross-communication seems to have only become more polarized. Not having any terminal value or end-goal, such as to ensure representation for the weak, or to provide ideal conditions for rational critical debate to take place, or instrumental values to ensure pluralism, social media platforms have, unsurprisingly, turned into unstructured and fractured communication spaces with no strategy to bring diverse voices together for meaningful deliberation that can help society achieve complex social goals. Cass Sunstein describes in *#Republic: Divided Democracy in the Age of Social Media*, of how in the United States the societal fragmentation, polarization, and radicalization produced by social media platforms has disrupted democracy itself (C. Sunstein, 2018). Rather than serving as vehicles for deliberation and policy formation, the rapid growth in the adoption of social media platforms, and the consequent economic and political power acquired by them, has in fact prevented their own effective regulation from being in line with societal values that can be imposed through the institutions of democratic governance (Zuboff, 2018).

Clearly neither the mass media nor the dominant social media platforms of today have been successful in creating a public sphere that can serve as a medium for deliberation. The process of deliberation that involves coming to a common understanding through bargaining, mediation, and eventually consensus through preference transformation, especially across social cleavages, and free of the influence of elites, appears to be a difficult ideal to achieve (Carpini et al., 2004; Elstub, 2006).

9.2 Building Blocks for Successful Deliberation

Given the challenges with building participatory societal communication platforms for successful deliberation and democracy, I next outline two related theories that help understand what should be the building blocks for public spheres to conduct rational critical debate. These theories help highlight the importance of plurality as an instrumental value for these meta-social good platforms.

9.2.1 Conversation Theory

Gordon Pask created conversation theory to explain how a pair of humans engaged in a dyadic conversation arrive at a common understanding on a given topic of discussion (Pask, 1976). He used the term *m-individual* (m-mechanical) to denote a human or host who exchanges information with another m-individual, and both then run certain procedures to operate on this information. An m-individual may contain multiple *p-individuals* (p-psychological) which are essentially stable concepts or memories that form as a result of the information procedures. When new information is received, the procedures evaluate it against existing p-individuals, test hypotheses, and possibly use it to update existing p-individuals or create new ones. The use of the term *individual* to identify memories, ideas, concepts, cultures, etc., is thus useful to give them an autonomy and independent existence.

For a successful conversation, the communicating entities must have a shared context or model of reality, which is a p-individual too. A conversation needs to first start with establishing a shared context, and then evaluate new concepts within this context. A single m-individual may host multiple mutually contradictory p-individuals, and a conversation can be successful as long as at least any one of these p-individuals is consistent with that of the other communicating entity, so that the information being exchanged can be understood by both the entities and then subjected independently to hypotheses testing. A deliberation between two people with opposed ideologies can have a chance of reaching a consensus if both these people have equivalent p-individuals to understand each other, and can then pose arguments to evaluate conflicting p-individuals and choose the one which is superior. The p-individuals can thus be considered as spanning multiple m-individuals in the case of collective thought or ideologies; culture can be reproduced through them identically or with modifications, and they can build affiliations with other p-individuals to strengthen their legitimacy. Habermas' concept of the lifeworld can also be modelled as a collection of p-individuals constructed through human thought and cultural experience.

Power relationships, hidden agendas, noise, emotion, and rhetoric are actively used in the real world to create and manipulate p-individuals (Navarro, 2001). Pask tried to model this in a network version of conversation theory, which he called the Interaction of Actors Theory (IAT), but died before he could complete the theory (Zeeuw, 2001). IAT introduces concepts similar to those in Actor Network Theory (ANT) and the theorization of power discussed in Chapter 6. *Coordinators* in IAT are actors who bring multiple m-individuals together (similar to the *enrolment* of allies in ANT) and can become sites of concentration of power to control who converses with whom. *Collectives* in IAT are networks of actors that have a unifying ideology (similar to *black-boxes* in ANT) and hence the actors in a collective are able to converse more successfully with one another. This strengthens their *associational power* and the ability to leverage it to gain higher levels of power, which can further influence others and continually expand the acceptance of their ideology. Evolving concepts that are not stable p-individuals as yet (have not becomes *immutable mobiles* of ANT) take shape through conversations, but their discoverability is controlled by network structures mediated by coordinators who may not necessarily embrace plurality. Technologies to support conversations and discoverability of new concepts (if they become *obligatory points of passage* of ANT) thus become important, and can gain power of their own to shape networks on which successful conversations can take place. IAT therefore attempted to build a sociological model of networked communication that could explain how conversations between multiple actors can lead to a shared understanding. Combinations of coordinators, collectives, and communication technologies could be used to model the creation of ideologies among m-individuals, create a resistance to other ideologies, and impact the ability of society to arrive at a common understanding or consensus. Conversation theory and IAT are thus useful to understand how new knowledge and concepts are adopted by society, or how they come into conflict with other concepts, and shape further formation of new concepts.

9.2.2 Information Evolution and Usefulness

With co-authors Jie Zhang and Robin Cohen, I introduced the terms of *context*, *completeness*, and *credibility* to describe three properties of information which make it useful for people (Seth & Zhang, 2008; Seth et al., 2015). I next relate these to the framework of conversation theory and go beyond to also explain the sociological processes through which these properties are realized. I will also show that plurality is essential for information to gain these properties. Societal participatory communication platforms, which intermediate the flow of information between people, need plurality as an essential instrumental value to facilitate actors from across social boundaries to come together and engage in conversations. This can help ensure that diverse concepts and ideologies can fairly interact and learn from one another.

We hypothesized that information is more useful for people when it is contextual, that is, when it is stated in terms of references that a person is familiar with, when its relevance is explained to the person, and it is expressed in the language of the person. We further hypothesized that information gains context when people similar to the recipient augment the information with re-articulations of the message, such as through comments shared on social media platforms. Since people are embedded in clustered social networks, where those in the same cluster or social network neighbourhood are likely to be similar to one another (McPherson et al., 2001), termed *homophily*, enhancements to the information by people in a social network neighbourhood are likely to make it more contextual for others in the neighbourhood. This can be directly related to conversation theory: People in the same social network neighbourhood are likely to contain p-individuals representing their shared context; these p-individuals would make it easier for the people to communicate with one another and help understand new information.

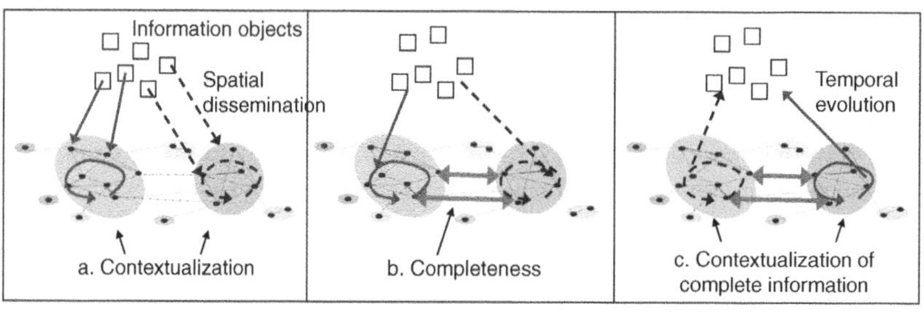

Fig. 9.1. Temporal Evolution of Information as it Moves Through a Social Network and is Made More Useful Through Contributions by Participating Users (Seth, 2008).

Completeness of information on the other hand is related to the comprehensiveness of its coverage of different viewpoints or aspects. We hypothesized that information gains completeness when people different from the original producer augment it with their own viewpoints, knowledge, and experience. Relating this to social networks, *weak ties* that extend to people in different social network neighbourhoods, or people who can simultaneously occupy different neighbourhoods, are likely to add completeness (Granovetter, 1973). Such people would host multiple p-individuals which help them understand different contexts, and can add new dimensions to the information by examining it from multiple contexts. We used these concepts to build a model of information evolution when messages travel through a social network, as shown in Fig. 9.1. As messages cross social network neighbourhood boundaries, contributions by people in these new neighbourhoods adds completeness to the message, and as more complete messages enter or re-enter neighbourhoods they gain context through contributions by people in the neighbourhood. The net effect thus is an improvement in the context and completeness of the message depending upon who participates in the conversations. Plurality can, therefore, be considered as imperative to facilitate learning through the creation of complete views.

Finally, we introduced the concept of perceived credibility of information as a personalized model of individuals about how they respond to new ideas. Different people may need the information to be less or more contextual before they can process it, similarly some people may have more closed views than others to accept information that is different from their own beliefs. People may also consider other factors of credibility such as the prior reputation of the original author(s), or endorsements by other reputed people, or the authoritativeness of the profession of the author(s), etc (Fogg, 1988). We used this model of credibility to evaluate the design of recommendation systems for participatory media messages. The relative social network positions of the recipient, the message producer, and other contributors, were used to determine the degree of context, completeness, and credibility of the messages, and accordingly recommend them to individual users.

A validation on social media platforms of this hypothesis that people have a personal preference towards what they perceive as credible, made us raise a significant concern, and which has been increasingly highlighted in recent times. Personalized algorithms for information recommendation should not just serve content in accordance with the discovered user preferences, but they should also encode some normative criteria for plurality, such as diversity or completeness, which can even overrule user preferences in case a user only prefers content that aligns with the user's viewpoint (Seth, 2008). More recent work has corroborated the transmission dynamics on social networks by showing that diverse information is more likely to flow through weak ties, and this can be used to create information recommendation systems that specifically target diversity rather than only reinforce user preferences (Bakshy et al., 2012; Matakos et al., 2020).

As building blocks for deliberation, such models of conversation theory or information evolution on social networks clearly show that participatory societal communication platforms need to consider several factors to create spaces for successful deliberation: they need to acknowledge the heterogeneity of ideologies and viewpoints in society, build strategies for plurality by fostering conversations between opposed groups of people by facilitating the creation of shared contexts, be sensitive to non-informational determinants of credibility such as the authority or elite status of the participants, and ensure that technology access or algorithmic artefacts, such as for information ranking and recommendation, do not create any inequalities in the discovery of diverse views. Fostering plurality, therefore, should be a core instrumental value for participatory societal communication platforms.

9.2.3 Federated Platforms

These ideas of upholding plurality in conversations on social networks are in line with Habermas' views on "conducting deliberation". To conduct deliberation in the presence of significant heterogeneity of viewpoints in society, Habermas suggests a layered model of informal federated public spheres which aggregate information into formal public spheres (Habermas, 1996). At the first layer, bottom-up public spheres that emerge from the lifeworld and follow usage norms of communicative action facilitated by civil society can serve as spaces for ethical discourse aimed at seeking clarification and understanding. Such spaces would enhance the context of the discussion for the local community, should be open for participation by all members, and should assist them in building a shared understanding of the discussion topic. At the second layer, multiple such public spheres should connect and engage with one another to build completeness, ignoring social hierarchies, and recognizing the plurality of views and issues represented in the different spheres. This will produce collective understanding and cultural memories with an emancipatory intent by mainstreaming voices of the marginalized. At the third layer, various issues that have emerged from the lower layers should be taken up in a formal public sphere such as a legislature. The formal public sphere should have structured protocols for exchange to avoid manipulation by private interests, and should have the goal of producing formal descriptions that can be free from risks of misinterpretation. These formal descriptions would then form appropriate laws for governance.

Given the role of conversations in enhancing the context and completeness of information, technologies for broadcast media are not suitable to build bottom-up public spheres because these technologies are not interactive (Salter, 2007). These technologies have also been enhanced over the years for deployment in a centralized manner which makes them affordable only to large corporations or the state, rather than for bottom-up public spheres operated by the civil society. The Internet, on the other hand, was developed as a platform specifically meant for interactive two-way communication, and can be used in decentralized ways by local civil society networks. Although current commonly available tools and technologies on the Internet have grown increasing centralized due to the emergence of monopolistic business practices and business models that benefit from centralization, fortunately, structures and protocols for low cost decentralized networks have also managed to sustain themselves.

A noteworthy attempt was made to realize Habermas' vision of a three-layered public sphere using the Internet. A global federated network of 70+ Independent Media Centres (IMCs) around the world, called Indymedia, was formed in 1999 to promote

> social, environmental and economic justice; assist the distribution of intellectual, scientific, literary, social, artistic, creative, human rights and cultural expression; illuminate and analyse local and global issues that impact ecosystems, communities and individuals; identify and create positive models for a sustainable and equitable society; and aid in a revolutionary social transformation of society that prioritizes people before profit. (Kidd, 2009; Salter, 2007)

These IMCs ran discussion forums on topics of globalization, trade, and indigenous cultures, following an etiquette of a commitment to active cooperation, disciplined speaking and listening, and respect for the contributions of every member. If a concern was fully discussed but not resolved, the disagreement was formally documented with explanations. Only after an aggregation of the discussions emerging from local forums could a proposal and its contestations be put up for vote by various representatives at the layer of the formal public sphere. This created an environment where conflict was not repressed and disagreement could be expressed without fear. The federated structure of IMCs thus institutionalized the principle of plurality, and was able to sustain the diverse lifeworlds constituting the network.

9.2.4 Learning From Small Successes

Several participatory communication and information discovery experiments have been conducted with the same ideal for plurality. A browser extension called NewsCube helped users visualize different aspects related to a news topic and to browse stories about these aspects, to help users counter media bias (Park et al., 2009). On similar lines, the Balancer browser widget locally tracked a users' browsing history and showed an indicator of whether the user had read more news articles related to one viewpoint and not another (Munson et al., 2013). In our own work, my students have built news aggregators that generate a news feed that is more balanced in its coverage of different aspects and framings than similar news feeds generated by tools like Google Alerts (A. Sen, 2021). We have also used similar principles to build a media bias monitor that observes different forms of bias in the coverage of current topics in mainstream English newspapers in India (A. Sen, 2021). Similar studies are rare for other mass media channels such as TV news, especially in Indian regional languages, where problems of biased coverage and even fake news are more prevalent. Checks and balances on the media through monitoring, and countering the bias by relying on news aggregators and tools to discover other viewpoints, are critical elements to build a well-informed public sphere.

Consuming diverse news alone is however not sufficient and needs to be supplemented with conversations between people, to make the information more actionable, and with greater context and completeness. Club 2.0 in Austria was a successful attempt at deliberation termed by Fuchs as *slow media*, which holds debates in groups of four to eight people, hosted by a non-expert, and telecast over television, to serve as a means for people to deeply understand an issue (C. Fuchs, 2021b). Similar methods operating at larger scales, called deliberative polls, have also been tried in some countries (Carpini et al., 2004). Social media platforms such as Facebook and Twitter are however not designed to specifically encourage civil dialogue, and further the opaque algorithms powering their recommendation system do not ensure plurality (Bakshy et al., 2015; Seth, 2020e). The moderation processes on these platforms are also opaque and centralized, which makes them less responsive to contextual aspects and to build usage norms that are conducive for deliberation (Seth, 2020e). Further, a recent experiment on Twitter showed that merely exposing users to opposing views did not help them empathize with these views, and rather simple prompts nudging users to reflect on their own personal experiences with friends who held different views increased their ability to understand others (Saveski et al., 2021). A different approach to participatory media is clearly needed. In this context, the content curation practices on Mobile Vaani (MV) may have some learnings to offer. MV has a federated structure that enables different communities to build their own norms and topic focus within which greater context can be facilitated, while sharing content across communities helps improve the completeness of the information. This cross-shared content is augmented with narratives recorded by our team that provide the relevant contextual background to help users appreciate these views. Additional policies include editorial guidelines followed by the moderators to pay attention to the tone in which discussions take place to keep the dialogue civil and mutually respectful, journalistic guidelines followed by the volunteers in reaching out to different stakeholders to bring out their viewpoints for a more complete coverage, and future plans to put in place distributed moderation processes through which users and volunteers will be able to participate in the moderation process (Seth, 2020e).

These examples provide a rich set of principles based on which societal participatory communication platforms can be built for deliberation with plurality as a core instrumental value. These principles include: having appropriate moderation guidelines, facilitated by technology, with a layered federated structure, which promotes context and completeness in information, is sensitive to notions of perceived credibility, fosters conversations on current affairs and long-term social and economic processes, and eventually contributes to an improved understanding of different topics by people. Building such platforms is crucial for society to empathize with pluralist worlds within it, discover their own morality and values, translate them into values that can be coded into law and policy, and thereby enable deliberative democracy, and social control over technology. Sustaining these communication spaces is not easy, though, since they face competition from other communication systems that favour existing exploitative and manipulative hegemonic structures. Many IMCs, for example, faced coercive action by the state against their operations, with frequent take-down orders and server shutdowns

(Salter, 2007). Social media platforms with traceable communication are known to help dictators spot dissenting targets easily, making it is harder for activists to coordinate resistance against them (Chenoweth, 2016). Platforms such as MV find it difficult to raise investment because of their specific social focus, as compared to other platforms which might be using similar technologies but align themselves with standard neoliberal economic goals. Access to limited financial resources makes it harder for independent media platforms to counter compromised public spheres, which tend to be well funded, addictive for the users, focussed on sensationalizing news rather than foster critical thinking and reflection, and also often backed by the state to specifically oppose local radical public spheres. Participation on bottom-up emancipatory platforms also suffers when people are not free to give time from their other needs, especially with having to undertake wage labour to meet their essential expenses. Despite these challenges though, social movements such as the IMCs, and sustained use of MV by its users for almost a decade, shows that communities recognize their value and that it is indeed feasible to build bottom-up public spheres that serve as forums for rational critical debate. More such platforms, and ideally working in a unified manner that is formally connected to law making, have the potential of becoming meta-social good projects through which democracy can operate. In the following section, I discuss the next step of political participation, once suitable communication and information platforms do exist: how might people use them to bring about institutional change and strengthen democracy?

9.3 From Deliberation to Public Action

Meta-social good projects for participatory societal communication can help people arrive at terminal and instrumental values that define their understanding of social good. This communication can form the basis for collective action by citizens through elections and other democratic processes to overturn hegemonic power structures (Dijk, 1989). Under what circumstances would this transformation of the existing systems happen? How would the values for social good be adopted by markets, or by the state to build regulations, and how can the state be held accountable to ensure that society's preferences are upheld? I start with presenting evidence about the effect of media communication on democratic processes, and then follow with examples of meta-social good projects that have brought about this transformation, albeit at small scales.

9.3.1 Evidence From Media Communications Research

Media effects research over the last century, especially in the United States, has led to many theoretical models to understand how media exposure affects public action and political activity. It was realized that the early *hypodermic syringe* models of people getting influenced based on what they read or heard, needed to be more complex because the opinions of people were also shaped by expressions and endorsements by opinion leaders, the existing opinions of people, and their propensity to get influenced (McQuail, 1981). This led to the *minimal effects* era in communications research where media alone was not believed to be a significant factor in shaping opinions and political activity, such as voting behaviour (J. Klapper, 1957). A reversal, however, happened with new research on agenda setting and framing which argued that the route of influence through media lay in its ability to shape what issues to think about and how to think, which led people to align more or less readily with different political views that they encountered (McCombs & Shaw, 1972; Scheufele & Tewksbury, 2007). Media effects were, therefore, not "minimal". Media rather served as a catalyst or filter for people to get influenced by other forms of communication and persuasion.

More recently, a new era of minimal effects is said to have come about due to the wide diversity of programming that digital technologies available for people to choose from (Bennett & Iyengar, 2008). Since people can rapidly switch to different programmes, media content, rather than being driven by editorial or political agenda, is now driven by consumer choice so as to not lose audiences. This leads media to simply echo existing preferences of people and to intensify polarization. Further, among people with no clear preferences or political interest, a disengagement from news and politics is being noticed since any accidental news listenership that used to happen earlier, when not enough cable channels or Internet-based programmes were available, has now been almost eliminated (Prior, 2001). On the other hand, it has been argued that the effects of media in the new era should not be considered as minimal because polarization on the one side and disengagement on the other, are both detrimental to building an informed public opinion, engaging in public affairs, participating in deliberative activities, and being responsive to the outcomes of deliberation (Holbert et al., 2010). With billions of users now on social media platforms, consuming information prioritized by algorithms that create echo chambers and filter bubbles, and which are not countered with adequate moderation policies, the effects of digital media are likely to be quite strong. This is probably evident too with faster appropriation of social media platforms by right-wing proponents being coincident with the rising wave of right-wing populism around the world (Funke et al., 2021; Nikolov et al., 2021).

On the upside though, self-expression on participatory media platforms has produced reflexive effects. People tend to participate more on online forums when they have greater exposure to news, and while this self-expression in some cases does lead to a hardening of their political stance, more often it brings about positive effects through self-reflection and learning (G. King et al., 2017; Valkenburg, 2017). This is because when communicating on public platforms, people become more committed to meaningful discourse, and asynchronous commentary helps improve the thoughtfulness in their responses (Cho et al., 2018). The positive effects of greater learning and being able to understand issues in depth, may outweigh the negative effects of polarization.

Greater online participation also leads to an increase in offline political and civic activity such as volunteering (Shah et al., 2007). This may happen as a result of greater self-reflection as discussed above, or by coming to know more easily of relevant campaigns in which people can participate, or even through social pressure, as demonstrated in a large experiment by Facebook to increase voter turnout (R. M. Bond et al., 2012; Rojas & Puig-i-Abril, 2009).

Such extensive and nuanced evidence over many decades of the effects of media on public action and democracy highlights the importance of media and participatory communication platforms to help people understand diverse views, deliberate with one another, form a public opinion, and follow through with political activities or civic participation. However, it is also evident that the platforms for such mass and participatory communication cannot be left to themselves. Careful management is needed to amplify their positive effects and minimize the negative ones. There is growing consensus that societal participatory communication and information platforms need to espouse normative goals to reduce polarization, bring about a pluralist understanding among people, strengthen democracy, and foster uses that encourage people to participate in civic life and build empathy towards others (Shah et al., 2017). These goals have specifically been explored among community media projects for development, to which I turn next as examples of meta-social good projects that have been able to successfully link information and communication processes with public action, and thereon to bring about changes in law and policy. These may provide hints on how to build more such transformational meta-social good projects in the future to incorporate values of social good accepted by society into the institutions of governance and economy.

9.3.2 Community Media

Community media aims to empower communities to create and share their own media, with the view that communities have diverse information norms and needs which are not easily met by media platforms controlled by outsiders (Howley, 2010). Community radio is among the most popular technologies for community media, where communities set up radio transmitters and create their own audio content on topics of relevance for local consumption. These topics may range from local economic activities such as agricultural advisory in rural areas, to discussions on issues of local governance, current affairs, and culture and entertainment, among others. Such community media projects come closest to helping understand how societal participatory communication platforms can lead to concrete public action and legislative changes (Clementia Rodriguez, 2011). These media platforms are not restricted to just serving as public spheres for pluralist information sharing: activist networks on the ground are able to leverage them to improve social accountability, community-based monitoring of public services, and build social movements (Lim et al., 2018; C. Rodriguez et al., 2014). These relationships between offline public action and online communication platforms can be complex and I draw heavily upon our own work with MV to demonstrate this.

Community media has traditionally been modelled along four theoretical approaches (Carpentier et al., 2003).

1. As a media-oriented intervention to serve the community. The community participates as planners, producers, and performers, to create content that is useful for the community. This could take the form of educational content for children, or agricultural content for farmers, or folk songs as cultural expression.
2. As an alternative to mainstream media. The community media then tends to have an agenda to reject commercial motives, assert human rights and social justice, encourage voluntary engagement, and critically analyze social and economic systems.
3. As a development intervention led by community media organizations. Such media tend to follow priorities dictated by donors in the social sector which may sometimes not be precisely or comprehensively aligned with the actual needs of the community.
4. Drawing from the metaphor of the *rhizome*, as a media that is fluid and eludes a specific definition. Opportunistically it may divide to take a form of any of the other three models, or co-exist simultaneously under multiple identities. Such media also then manages to serve a networking function because of its multiple touch-points with the market, state, civil society, and community. Although this has been argued to be a source of weakness that prevents rhizomatic community media from being able to challenge exploitative and dominating power structures of the market and state, such community media is often more resilient and longer surviving than counterparts that are rigid to one particular form (Bosch, 2010). Our own experience from MV, which, as I discuss next as a rhizomatic form of community media, also indicates that this networking function is crucial in bringing about several kinds of meta-social good impacts that may not have been possible had MV taken an altogether antagonistic stance to the state or market, or an entirely donor-driven agenda.

MV simultaneously straddles many different definitions which often makes it hard to understand. One, it plays a strong community-supporting role, which, to a large extent, is driven by the community itself, to formulate different programmes in different geographies based on the specific needs or demands of the community members. Topics such as local news, agricultural advisory, career counselling, children's education, social entitlements, and labour rights especially for informal sector workers, emerged organically and were built into popular programmes directed by community volunteers (Moitra et al., 2018).

Two, MV raises funds from donors to run programmes aimed at specific social development goals (D. Chakraborty et al., 2019). The regular user base that may call MV for local news also actively consume these development programmes, on topics such as health and nutrition of pregnant mothers and small children, early marriage, domestic violence, women entrepreneurship, and other topics. Similar to topics that have emerged bottom-up, the content for these externally imposed topics is created in consultation with the community and reflects the local context, even though the need for these programmes may not have been originally expressed by the community members. In all cases, however, these topics were subsequently acknowledged by the communities as being important and relevant for their lives and saw active participation of the community.

Three, MV often nurtures discussions on current affairs and policies that may bring out many oppositional views by the users towards the actions of the market or state. The Gram Vaani team aggregates these responses and helps represent the views of the community to the government by writing letters and meeting government officials. At the same time, however, government departments use MV to make announcements that can reach remote communities that are not easily accessible through other media channels. This networking role played by MV is crucial in bringing the state and citizens closer together. An area where this becomes quite significant is with escalation of citizen grievances related to social protection schemes (D. Chakraborty et al., 2017; Gupta et al., 2021). People call MV and report problems they are facing with government schemes, which the community volunteers try to resolve by assisting them, failing which they escalate the issue to senior government officials. During the COVID-19 lockdown in India, the MV platforms were heavily used to serve this role of social accountability to improve access of social protections for citizens who suffered from drastic income loss due to unemployment and economic slowdown (Wang et al., 2021). The local MV communication platforms were therefore instrumental in enabling the offline network of volunteers and activists to hear about problems that people were facing, and to then use the public nature of the platform to hold local government officials accountable to bring about quick action. Rather than building an antagonistic relationship with the government, however, an agonistic relationship was created, where the volunteers after having understood the underlying implementation problems, also helped the officials resolve some of these issues through local action.

Four, MV creates strong community ownership through a careful alignment of incentives for voluntary participation by the community members (Moitra et al., 2018). Collectives have otherwise faced free rider effects that dampen the motivation of volunteers to keep contributing their time and resources while others benefit without putting in equivalent effort (Olson, 1965). Clarity on the objectives of the collective, of whether it is meant for social development or for other purposes, can also be a cause of conflict among volunteers to align their individual motivations towards a common goal (Butler et al., 2007). Better communication and observability of others, possible now with online forums, can assist with the running of collectives in the contemporary technological context (Lupia & Sen, 2003; Ostrom, 1990). Such a conducive environment was created within the MV volunteer clubs, where regular meetings between the volunteers and exchange of information about activities conducted by individual volunteers helped create mutual accountability. Further, strong solidarity effects emerged from teamwork by the volunteers to collaboratively solve community problems. Dozens of local campaigns on social welfare programmes were initiated by the clubs such as on the quality of food served to school children for their mid-day meals, the practice of caste-based segregation in the seating of students in schools, regularization of opening times for fair price shops, and other campaigns, where the volunteers came together and built organized strategies to gather evidence and draw attention of the administration to these issues (Moitra et al., 2018). This created strong bottom-up ownership since communities were able to define the agendas of their local platforms and sustain their involvement in running the platforms, with minimal external coordination facilitated by Gram Vaani. The offline volunteer networks have thus been instrumental in creating the local MV communication platforms into vibrant public spheres that discuss relevant topics grounded in the needs of the communities, and persuade institutional action on these topics.

All the elements of community ownership and community-driven agenda, donor agenda, accountability of the state, and use of the medium by the state, are therefore able to form a symbiotic relationship with one another in MV. Typically the donor agenda provides funds that are also able to subsidize other functions, while these functions bring strong social credibility and community acceptability of MV, which, in turn benefits donor outreach. As a federated network of local public spheres, MV is thus able to balance multiple priorities at the local level, while also aggregating the community needs to represent them to the government at the local as well as the state and national levels. A theory of change inferred through the Most Significant Change technique of interviewing MV users, shows these links clearly (Moitra et al., 2019). The choice of voice-based technology improves accessibility and inclusion. Management processes of moderation foster usage norms to create safe spaces for self-expression. Editorial credibility gained through fair treatment of these self-expressions encourages further participation. The context, completeness, and credibility of the information shared on the platforms create useful learning spaces. Support by offline volunteer networks helps publicize MV among marginalized communities and recognize relevant topics to feature on MV. A multi-level engagement facilitated by Gram Vaani with various government and non-government stakeholders provides pathways to bring additional benefits to the community, especially when some of these relationships get institutionalized. Volunteers and activists are able to use MV platforms to hear what problems are faced by people, and hold appropriate stakeholders accountable for action. Agency effects are thus perceived when people are able to express themselves freely, hold local stakeholders accountable, network with other media and organizations, and even influence policy.

Similar agility in bringing strong community impact, as well as policy changes through local and federated public spheres, has been a feature of other community media projects as well around the world (Howley, 2010). They nurture ground-up discussions, which, if managed well, can serve as spaces of deliberation. They also build networks with the state through which communities can influence policy. Networks of multiple community media instances can even work in unison on common objectives to build a louder voice for citizen demands. The answer to creating societal participatory communication platforms may therefore very well lie here, to follow the three-layered structure suggested by Habermas, composed of a network of community media setups, and aggregated into formal structures that respect plurality and are able to connect with institutions of the state and markets.

But challenges do exist. Sustaining such alternate media platforms, and being able to follow an agenda of community empowerment, has obvious political economy challenges (C. Fuchs & Sandoval, 2015). Further, it can be difficult to sustain participation in such forums of active citizenship when many more distracting media channels and social media becomes accessible to people. However, building more such spaces based on the learning so far can contribute towards the vision of societal participatory communication platforms. These platforms are urgently needed to prevent further refeudalization of the lifeworld by individualism instead of collectivism, instrumental rationality instead of cooperative development and solidarity, and manipulated social opinion instead of genuine public opinion to shape laws in a democracy.

9.4 Summary

I have tried to show that local public spheres constructed as bottom-up community media, and formally or informally aggregated into campaigns that can influence regulation and policy-making, are viable means of creating societal participatory communication platforms that uphold the values of plurality and democracy. The use of appropriate technologies to run these platforms, based on the specific needs, constraints, and opportunities in different communities, can ensure inclusion. Communication primitives that follow guidelines such as context, completeness, and credibility can foster learning environments through which communities can understand themselves as well as others. This can bring empathy, attention to social justice, demand for social accountability, aspirations for better governance, convergence towards essential values for social good, and eventually, through representation at the policy-level, can bring institutional changes in the systems of the state and markets. An offline–online integration between the activist networks and communication platforms can provide mutual support to each other to amplify impact. Deliberation through conversations forms the bedrock for societal learning, and such societal participatory communication platforms are therefore necessary to challenge the hegemonic structures that have colonized the lifeworld and monopolized technology-design paradigms towards forms that cause disempowerment. Technologists need to work on such meta-social good projects in partnership with society and bring about a transformation in the current structures of the state and markets, for only such a transformation can bring more humanism.

Fostering such societal participation at scale can be done by building coalitions of civil society organizations, community-based institutions, labour unions, and rights-based activists, who are typically unified in terms of a set of shared values for equality. This can create the necessary *societal power* when multiple social movements are able to join hands to strengthen democracy and institutionalize values for social good (Dijk, 1989). India recently witnessed such massive collaboration between multiple social movements during farmer protests at the Delhi borders, which went on to dent the electoral results of the ruling political party in several state elections, and finally led to a repeal of the laws (T. Ali, 2021; S. Kumar, 2021; Sawhney, 2021). A well-connected public sphere, constituted of multiple small communication ecosystems, was essential to have people understand and acknowledge the issues with the new farm laws, and the government's brash attitude of dismissing consultations to subvert democracy. Similar movements in which technologists and society can join hands through participatory communication platforms are needed to impose society's control over technology, and to strengthen democracy itself.

Building such meta-social good initiatives is necessary to bring transformational changes in contemporary institutions and paradigms of technology development. Such initiatives can also become a medium for technologists to learn more about the impact of their technologies on society, and to inform society of technological risks. Further, platforms carrying bottom-up self-expression provide valuable opportunities through digital ethnography that can help technologists connect better with their users to learn more about them (Tacchi, 2012). Technologists can then collectivize to change the design and management processes in their organizations and governments to align them with essential values of social good. Irrespective of where in the technology production value chain they may be working, technologists have a range of methods that they can use, as described in Chapter 8, to influence their organizations and governments. This will push the world towards avoiding technologies that disempower, adopt paradigms for empowering technologies, undertake projects that are unanimously meant only for social good, and enable technologists to rediscover their humanism by being better connected with the rest of society and participate in the emancipation of the weak from the current structures of domination and exploitation.

Chapter 10

Conclusions

Nothing summarizes the key point in this book than Martin Luther King, Jr's speech in 1954 called *Rediscovering Lost Values* (King Jr, M.L. 1954). The trouble, King said, is not that

> we don't know enough, but it's as if we are not good enough. The trouble isn't so much that our scientific genius lags behind, but our moral genius lags behind... The real problem is that through our scientific genius we've made the world a neighbourhood, but through our moral and spiritual genius we've failed to make of it a brotherhood.

He argued that

> to go forward, we must go back and rediscover some mighty precious values that we've left behind. That's the only way that we would be able to make of our world a better world ....

We are divided on what are essential values that we should live by, and around which we should architect our systems and build technology. We seem to have, in fact, permitted the growth of systems that are value-less and proclaimed it as a useful feature. To make it worse, our morality, which should be the foundation of these systems, is actually getting eroded by them. The task in front of us as technologists is formidable: not only do we need to safeguard against technology being used to entrench these systems further in people's lives, but to use technology to displace them or transform them into new systems that have firm moral foundations. Marx saw morality as humanism that is constructed through social relationships of production. Freire saw this in terms of humanists to support the oppressed. Gramsci saw this as being led by organic intellectuals drawn from among community groups. Habermas saw the relevance of communicative participation in public spheres for societies to learn from communities and arrive at a consensus. Nino saw this deliberation as translating to law in a democracy. The

Technology and (Dis)Empowerment: A Call to Technologists, 183–185

Published under exclusive license by Emerald Publishing Limited
doi:10.1108/978-1-80382-393-520221010

answers have always been in front of us to create a new paradigm for technology that follows these principles, we now need to move in earnest in this direction.

I have argued that to build paradigms which can ensure that technology unanimously gets used for social good, we have to work on many aspects.

We need to train ourselves in the language of ethics to be able to reason how to design and manage technology projects which can avoid harm. Approaches of ethics by design are essential but alone not sufficient: rather the entire technology and project management architecture needs to be evaluated for consistency with underlying ethical principles.

We further need to acknowledge that the rationalist approach followed in the natural sciences and for technology development is not suitable to understand and govern technologies across diverse societies. This argument is not meant to negate the relevance of technology and science, it is rather meant to draw attention to the limits of the rationalist framework in understanding and controlling the interactions between technology and society.

A compliance-oriented approach through external regulation is also not sufficient. The entire organizational ethos needs to be aligned with common ethical foundations to deal with the uncertainty of outcomes that can arise from technological innovations.

What should be these underlying ethical foundations for technology for social good projects? The definition of social good itself needs to be democratically discovered through the participation of society on meta-social good participatory communication systems that support learning and deliberation. A definition of social good cannot be externally imposed. Societal learning is possible only by embracing universal values of plurality and diversity in a democracy, which can enable people to know what they do not know.

To guide action based on learning emerging from societal communication, requires another universal value of power-based equality to prevent existing stakeholders in positions of power to misuse their power to entrench themselves further. This is true at all levels, within organizations, where power relationships between workers and managers need to be equalized, and across organizations and institutions to form rules and laws that can guide them towards equality, as well as among the direct and indirect users of technology so that their mutual social relationships are also made more equal.

Power-based equality is however unlikely to come by without demanding it. It requires collective action by technologists within their organizations, to gain power through which they can keep a check on technology projects involving their labour.

It also requires external political activity by technologists as members of society, in partnership with citizens and consumers, at the local, national, and global levels, to make society aware of technology risks, and collaborate with society to ensure that through democratic institutions the values deemed socially good by society are built into the regulatory structures of the market and state, and consequently adopted by technology organizations.

Further, participatory communication systems are needed also for technologists to connect with their users so that they can, with empathy, understand the

impact of their technologies on diverse sets of users. This will let technologists understand how eventual use-value is derived from the output of their labour, decide whether it is in line with social good goals, and to overcome the social and economic divides that have emerged between them and the bulk of society. This will facilitate technologists to design and manage technologies by being embedded in the social systems where they are used. Only through this praxis will social good get defined and institutionalized, and more people will join the movement of using technology for social good.

An urgent change is indeed needed in the paradigms that currently guide technology development and scaling. Paradigms that emerge from the existing structures of the state and markets tend to increase inequality, reduce freedom and empowerment, and ultimately de-humanize the technologists behind these technologies. We have a long agenda in front of us to overcome these paradigms, and transform them into paradigms that produce technologies and projects to truly empower people and bind us together in positive social relationships. We need to do this not only because it is the right thing to do, but it also feeds our own humanism. The struggle for humanism in our work is one and the same as the struggle for humanism in society. It is up to us to architect a new system within which we want to work and innovate for social good.

Chapter 11

Epilogue

Let me discuss a few examples of how the ideas in this book have shaped project objectives and planning for a few projects at Gram Vaani, with my students, and with other collaborators. This may help demonstrate that the way we think influences what we build, and emphasize the need for new paradigms in which we should be thinking about technology.

COVID-19 vaccination scheduling: The first example is related to the COVID-19 pandemic. This section was written in May 2021, when India was devastated by the second COVID-19 wave that ravaged urban metropolitan centres. Hospitals ran out of beds and oxygen leading to many avoidable deaths, and the pandemic had spread to deep rural hinterlands where healthcare access was many times bleaker. With the virus mutating to more infectious forms, rapid vaccination was the most important tool to keep people safe. Notwithstanding the multiplicity of issues that India faced with the supply of vaccines, I draw attention here to further exclusionary mechanisms that emerged within India for people to book vaccination slots. A central web portal was created for slot bookings but which was usable only by people having smartphones and Internet access, a mere 50% of the population. Walk-in vaccinations were not allowed for age groups between 18 and 44, no IVR or call centre-based mechanisms were put up, and government-authorized Internet help desks were closed due to state-level lockdowns in most locations. Not only was the offline population invisible to the government, it was also invisible to most technologists, as soon became obvious. The government opened up Application Programming Interfaces (APIs) through which vaccine slot availability could be polled by third-party applications; many programmers jumped at this opportunity and released email and instant messaging-based alert systems to push notifications to people when vaccines became available at vaccination centres close to them. The contest for booking whatever limited vaccines were available thus turned even more unequal, between technology savvy people who could jump between multiple applications, and those with limited skills or resources who were not able to use this eclectic technology mix.

The Gram Vaani team saw the obvious exclusion that such systems were causing and launched a campaign with *online volunteers* to help people register for

Technology and (Dis)Empowerment: A Call to Technologists, 187–192

Published under exclusive license by Emerald Publishing Limited
doi:10.1108/978-1-80382-393-520221011

vaccines. The team also considered building an application for volunteers to systematize their work in assisting other people. We can see through this example that simply the visibility of different population groups to technologists can influence the problems they recognize and then choose to solve. A clarity on terminal goals aligned towards equality could have brought a more informed perspective to technologists.

Strengthening collectives: Recently I connected with two agriculture-focussed social development partner organizations to co-author funding proposals that use *AI for social good*. Both partners were focussed on supporting smallholder farmers by organizing them into collectives but were trying to solve different problems in this space. The first organization was trying to improve internal coordination among members of the collective, and would benefit from tools that helped with crop planning, yield estimation, water management, and internal communication. Many technological innovations in price forecasting, remote sensing, communication, and question-answering, could thus be useful. The framing for social good in this case was, therefore, to put technology to use by the collective so that it could improve the productivity of its member farmers.

The second organization was engaged in helping several collectives to sell their produce at the best price, by coordinating its procurement from the farmers, transporting it to markets, and selling it to vertically integrated relationships it had built with food processors and retailers downstream in the agricultural value chain. The challenge the collectives faced was from corporations who had large working capital and were therefore able to offer higher procurement prices to farmers, and offset this loss by simultaneously procuring surplus produce from earlier seasons which had been purchased by the government and was subsequently re-sold at low prices in reverse auctions. Thus, by using their surplus capital, corporations could systematically wipe out competition from the collectives, village by village, and acquire a monopoly position at which time they could drive down the prices they gave to farmers. Our collaborator organization described their need as requiring a market intelligence system that could help the collectives to compete more effectively against such large predatory corporations. Such a system would consist of tools for price forecasting, a price surveillance network, and risk analysis for pricing in different scenarios based on various actions that could be taken by the competing corporations. The partner was confident that these tools could help them compete more effectively, and further, if the collectives were able to transparently communicate to member farmers about the prices at which crops were bought and sold, it would even strengthen the credibility of the collectives as ethical market entities so that the farmers themselves would not sell to unethical corporations even if they were offered a higher price. The framing in this case clearly was that of a meta-social good project to counter the dominating tendencies of capitalist market players.

While both the proposals made innovative use of technology for social good, the first one operated within the existing market system and aimed to strengthen the capability of collectives to participate in the system, while the second one aimed to change the system by neutralizing power-based inequalities in the market so that farmers and farmer collectives in the long run could get better prices for their work. This example shows that different underlying ideologies and values can lead to the identification of different problems that are worthy of being solved.

Supporting youth: At Gram Vaani we have been thinking for a long time of building a new programme focussed exclusively on youth in rural areas. Based on our extensive interactions, it became clear to us that young people had the energy and ambition to dream big, but had little exposure on how to go about it. They did not have access to information and knowledgeable mentors or role models, and were often at a loss of what subjects to choose for specialization in their education, or how to search for a job, or prepare for interviews. As we discussed and designed what services should we build for this programme, we realized that we would often struggle between two entirely distinct value systems. One was individualistic, focussed on providing resources and counselling to individual youth with the clear proposition that the services would help them get ahead in life. This led us to follow-on needs to connect the youth with skilling organizations and job placement platforms. The other was a collective approach, focussed on linking them with mentors and peers, and with their wider community, which led us to conceptualize different activities altogether. These included ideas for the youth to leverage their superior technical skills to support their communities in access to social protection schemes, or vaccine registrations as described earlier, or experiment with new agricultural techniques and models. This example again shows that different underlying values lead to framing of different problem statements, and to create radically different worlds – individualistic and market based in one case, and collective and cooperative in the other case.

Monitoring of socio-economic development: A vision I am pursuing with my students at IIT Delhi is to build a Giant Economy Monitor that will bring disparate datasets together to monitor socio-economic development in India (ACT4D, 2019). In the absence of frequent population censuses and additional delays in the public release of data, we are using satellite imagery to infer the true state of socio-economic development (C. Bansal et al., 2020, 2021). This data, available over different years, and at fine spatial resolutions, allows us to monitor the development of different cities and villages over time, which we are now relating with factors such as social welfare expenditure, political alignment of elected representatives, and corporate linkages of the elite from these areas (C. Bansal et al., 2020). Further, we are using news articles, social media data, and eventually also participatory media data (sought through on-demand surveys), to infer qualitative explanations and social outcomes covered in the news and expressed by local communities. We are also applying similar techniques to identify illegal mining activities, forest clearance violations, and to understand the environmental impact of extractive development. Our goal is that such a platform can guide governments towards the equitable distribution of welfare expenditure, identify areas of exclusion and political or corporate corruption, highlight anomalies both positive and negative to help build a better understanding of development processes and their impact, and hold governments accountable to build equitable and just policies. This is therefore a meta-social good project to improve governance through data and analysis tools to instrument the development process and ensure that it is steered appropriately.

Agriculture stacks: There has been a recent flurry of activity around the world with building *technology stacks* in the agricultural space. The Linux Foundation has initiated the AgStack project which seems to be a suite of tools for supply-chain management, crop planning, equipment management, IoT integration, etc., but specifically mentions that it will not build applications and platforms itself (AgStack, 2021). Even if the tools are designed through participatory and consultative methods to ensure that they fill a relevant need, the absence of any clearly specified terminal values such as power-based equality implies that the tools are likely to be appropriated by entrenched entities such as corporations and large farmers, and will lead to a further increase in inequalities. As acknowledged by several collaborators, this absence of an underlying ethics-based foundation in the Free and Opensource Software (FOSS) movements has led to its ineffectiveness in being able to control how FOSS products are put to use in applications that disempower people rather than empower them. We are therefore discussing a narrower set of principles which we summarize as FREEDOM principles – *free*, *open*, and *responsibly* and *ethically designed* and *managed* digital technologies – that can identify ethically reasoned instrumental and terminal values, and strengthen the FOSS movement to do social good.

The Indian government also recently announced a consultation for the Agristack concept being designed as a digital public good, called India Digital Ecosystem of Agriculture (IDEA) (GoI, 2020a). It uses the same principle as Aadhaar of having a unique ID, in this case an ID for each farmer, and an ID for each farmland, around which tools can be built to provide social protection schemes, customized advisories to farmers, crop planning recommendations, market linkages, digital transaction platforms, and crop insurance. Such concepts focus exclusively on individualization instead of collectivization. They stand the risk of forcing simplistic categories of land–farmer relationships which do not account for complex land tenancy and sharecropping relationships that actually exist in practice. They are likely to introduce new problems for farmers due to poor socio-technical interface management for data collection at the last mile, as has been clearly identified with problems arising in systems built on Aadhaar (Gupta et al., 2021). They open up possibilities for misuse of farm data, as is known from digital land grabbing experiences when land records are digitized and accessible to markets (Benjamin et al., 2007; GRAIN, 2020). The existence of data without clearly defined purpose limitations also opens up new spaces for surveillance capitalism, to learn to predict farmer behaviour and benefit from it, and eventually even manipulate it (Zuboff 2018). The project therefore seems to follow the same motivations as Aadhaar and similar technology platforms, especially those projected as digital public goods, which actually have an aim to create new markets for the deployment of formal capital once data and transactions are digitized. This is followed by financialization with further deployment of money capital, all of which eventually leads to dispossession and exploitation. The promise of improving farmer livelihoods through digital technologies deployed in the Agristack architecture may therefore just be a ruse to bring agriculture into the market system. Enabling access to social protection layered on Aadhaar may similarly just be a means for rapid user acquisition since large farmer registries now already exist due to the mandated use of Aadhaar in welfare schemes. This brings together all the aspects I have discussed so far to do with an ethics-based design and management of the technology platform to ensure that it leads to social good, along with the need for meta-social good projects to control the political economy around the technology. If the terminal end-goal is indeed to benefit farmers to get a good price for their produce, to cope more easily with crop damage, and to improve farm productivity, then other architectures are more suitable, such as architectures to strengthen farmer collectives and provide tools to them without the need to track each farm or farmer individually. If such individual data is indeed useful then approaches such as the digital commons for collectivized ownership of the data, with sufficient and effective data protection laws, and privacy enhancing technologies, can, instead, be adopted. A larger question we are also trying to answer is whether such productivity-enhancing technologies are really the need of the hour, or should we rather focus on other technologies which counter power-based inequalities in the largely informal agricultural production and marketing value chains in India, or if technologies can even address these issues instead of forming better policies for price support, irrigation, and market and storage infrastructure (Madaan et al., 2019).

Education: Finally, this discussion of changing the current paradigms of technology design and management through a praxis by technologists also raises the question that how should the education curricula be changed so that students who grow up to be technologists, take care of producing and using it in ways that empower people. Recalling my own experience in India, I feel that many crucial elements needed to nurture such a praxis seem to be missing at both the school and higher education stages. First, the teaching of history can help students understand the political economy and how it has continuously shaped technology and innovation, but history is more often taught as memorizing a list of events. Second, reductive methods are generally followed for the teaching of engineering and the natural sciences, which misses on providing a "big picture" understanding of how these laws and systems interact with one another. Systems thinking, which, today is considered as a niche discipline in itself, should be brought out as a core topic for the teaching of engineering, design, and science. Third, literature which is mostly taught to build language and comprehension skills, can be used to highlight the importance of qualitative narratives to understand the realities of people living around the world under different social systems at different points of time. Stories are key to understanding the relationship between abstract theories and their actual realizations. Narratives can also more easily evoke questions of morality than abstract theories, and build an ethical basis to thinking about right and wrong, as well as the humility to recognize that perfect solutions may be impossible to find in many circumstances. Fourth, the recent push towards teaching computational thinking in schools needs to emphasize on the loss of precision that comes with informatization of the complex analogue real world. This will help students appreciate the imperfection of computational technologies to – especially – model social processes, and not get carried away with the successes seen with the modelling of natural processes. Fifth, the topic of civics is often taught as memorizing a set of procedural rules according to which government operates, and needs to shift to explaining the reason behind these rules – of respecting values of freedom, autonomy, rights, consensus, and accountability, so that students can understand the spirit of democracy. This can further be used to convey that these rules are not perfect, they need to be continuously improved in democratic ways. Sixth, an appreciation of various imperfections in governance, science and technology, computing, and laws, can be used to highlight the importance of individual and collective responsibility of technologists. This is critical to form a strong foundation for students to take pride in human intellect and ingenuity, but also have the humility to recognize gaps and deficiencies that remain. Finally, this realization for an ethic of responsibility should go further to addressing social problems that we face today, so that technologists can use their skills to create meaningful use-value and build positive social relationships with the rest of society.

I have tried to demonstrate through these examples that we need new paradigms in how we conceptualize, design, build, manage, and scale technologies. The dominant paradigms embedded in the current systems of the state and markets are driven by values of individualization, competition, control, and hierarchy, and a belief in rationalist means to control and steer technologies and societies. This paradigm produces technologies that increase inequalities, reduce humanism in society, and serve largely to entrench the hegemony of the paradigm itself. New paradigms are needed that nurture values of collectivism, inclusion, support and care, pluralism, and power-based equality. These paradigms through democratic governance and participatory societal communication systems will lead to the discovery of new constellations of values acceptable to society, which can guide the effort of technologists towards building a different set of technologies that increase humanism in society. These values need to be discovered through democratic means, and our role as technologists is to build systems that can nurture this discovery process through critical reflection and learning in the public sphere and through democracy. Meta-social good systems that assist in upholding democracy are critical to bringing this social control over technology, since if society does unwittingly reject meaningful democracy itself and fails to counter the increasing inequality and corruption of democracy by the powerful, then society will lose control over the very definition of social good, and thereby forfeit any ability to ever control technology towards social good and responsible outcomes.

I have faith that if we build such systems for social good and meta-social good then humanity will be able to come around to solving the challenging global-scale problems that confront us today and to uphold humanism in society. It is our very humanism that is being challenged today and we need to come together to preserve it in the world.

Bibliography

Books

Anderson, E. (2017). *Private government: How employers rule our lives (and why we don't talk about it)*. Princeton University Press.

Babbage, C. (1832). *On the economy of machines and manufactures*. Cambridge University Press.

Barocas, S., Hardt, M., & Narayanan, A. (2019). *Fairness and machine learning: Limitations and opportunities*. http://www.fairmlbook.org

Beer, S. (1975). *Platform for change*. John Wiley and Sons.

Birkinbine, B. J. (2020). *Incorporating the digital commons: Corporate involvement in free and open source software*. University of Westminster Press.

Braverman, H. (1974). *Labor and monopoly capital: The degradation of work in the twentieth century*, Monthly Review Press.

Brian Arthur, W. (2009). *The nature of technology*. Penguin Books.

Castells, M. (1996). *The rise of the network society* (2nd ed.). Wiley-Blackwell.

Castoriadis, C. (1988). *From the workers' struggle against bureaucracy to revolution in the age of modern capitalism*. University of Minnesota Press.

Chatterjee, P. (2020). *I am the people: Reflections on popular sovereignty today*. Orient Blackswan.

Collingridge, D. (1980). *The social control of technology*. St Martin's Press.

Collins, D., Morduch, J., Rutherford, S., & Ruthven, O. (2009). *Portfolios of the poor: How the world's poor live on $2 a day?* Princeton University Press.

Cooley, M., & O'Grady, F. (2016). *Architect or bee? The human price of technology*. Spokesman Books.

Crouch, C. (2011). *The strange non-death of neo-liberalism*. Polity.

Dahlbom, B., & Mathiassen, L. (2016). *Computers in Context: The Philosophy and Practice of Systems Design*. Blackwell Publishers.

Duetsch, K. W. (1963). *The nerves of government: Models of political communication and control*. The Free Press.

Escobar, A. (1995). *Encountering development: The making and unmaking of the third world*. Princeton University Press.

Escobar, A. (2018). *Designs for the pluriverse: Radical interdependence, autonomy, and the making of worlds*. Duke University Press.

Foucault, M. (1984). *The Foucault Reader: Edited by Paul Rabinow*. Pantheon Books.

Freire, P. (1970). *Pedagogy of the oppressed*. The Continuum Publishing Company.

Fuchs, C. (2013). *Social media: A critical introduction*. Sage Publications Limited.

Gasper, D. (2004). *The ethics of development*. Edinburgh University Press.

Gorz, A. (1998). *Critique of economic reason*. Verso Books.

Graeber, D. (2015). *The Utopia of rules: On technology, stupidity, and the secret joys of bureaucracy*. Melville House.

Gramsci, A. (1971). *Selections from the prison notebooks*. Lawrence and Wishart.

Greenwood, D.J., & Levin, M. (2007). *Introduction to action research: Social research for social change*. Sage Publications.

Habermas, J. (1989). *The structural transformation of the public sphere: An inquiry into a category of bourgeois society*, Translated by Thomas Burger with the assistance of Frederick Lawrence. MIT Press.

Habermas, J. (1996). *Between facts and norms*, Translated by W. Rehg. London Polity Press.

Hardt, M., & Negri, A. (2000). *Empire*. Harvard University Press.

Harris, E. D. (2016). *Governance of dual-use technologies: Theory and practice*. American Academy of Arts and Sciences.

Harvey, D. (2014). *Seventeen contradictions and the end of capitalism*. Profile Books.

Hayek, F. A. (1944). *The road to serfdom*. Routledge Press.

Healy, M. (2020). *Marx and digital machines: Alienation, technology, capitalism*. University of Westminister Press.

Hensman, R. (2011). *Workers, unions and global capitalism: Lessons from India*. Tulika Books.

Herman, E. S., & Chomsky, N. (1988). *Manufacturing consent: The political economy of the mass media*. Pantheon Books.

Howley, K. (2010). In K. Howley (Ed.), *Understanding community media*. Sage Publications.

Illich, I. (1973). *Tools for conviviality*. Harper and Row.

Jansen, S.C., Pooley, J., & Taub-Pervizpour, L. (2010). In S. C. Jansen, J. Pooley, & L. Taub-Pervizpour (Eds.), *Media and Social Justice*. Palgrave Macmillan.

Jonas, H. (1985). *The imperative of responsibility: In search of an ethics for the technological age*. Chicago University Press.

Kanter, R M. (2010). *Super Corp: How Vanguard Companies create innovation, profit, growth and social good*. Crown Publishing Group.

Khera, R. (2019). *Dissent on Aadhaar: Big data meets big brother*. Orient Blackswan.

Kleiner, D. (2010). *The telekommunist manifesto*. Network Notebooks Series.

Kohli, A. (2012). *Poverty amid plenty in the New India*. Cambridge University Press.

Krishna, A. (2017). *The broken ladder: The paradox and the potential of India's one billion*. Penguin Random House.

Latour, B. (2005). *Reassembling the social: An introduction to actor-network theory*. Oxford University Press.

Lessig, L. (1999). *Code and other laws of cyberspace*. Basic Books.

Levy, S. (1984). *Hackers: Heroes of the computer revolution*. O'Reilly Media.

Lukes, S. (2004). *Power: A radical view*. Palgrave Macmillan.

MacKenzie, D., & Wajcman, J. (1999). In D. MacKenzie & J. Wajcman (Eds.), *The social shaping of technology*. Open University Press.

Marx, K. (1844). *Economic and philosophical manuscripts*. Progress Publishers. https://www.marxists.org/archive/marx/works/download/pdf/Economic-Philosophic-Manuscripts-1844.pdf

Marx, K. (1867). *Capital: A critique of political economy, Vol. I. The process of capitalist production*. Progress Publishers. https://www.marxists.org/archive/marx/works/download/pdf/Capital-Volume-I.pdf

McLuhan, M. (1964). *Understanding media: The extensions of man*. McGraw-Hill.

McQuail, D. (1981). *Communication models for the study of mass communication*. Longman House.

Meadows, D. H. (2008). *Thinking in systems: A primer*. Chelsea Green Publishing.

Mills, C. W. (1951). *White collar: The American middle class*. Oxford University Press.

Mills, C. W. (1956). *The power elite*. Oxford University Press.

Mourkogiannis, N. (2014). *Purpose: The starting point of great companies*. St. Martin's Press.

Mumford, E., & Weir, M. W. (1979). *Computer systems in work design: The ETHICS method: Effective Technical and Human Implementation of Computer Systems*. John Wiley and Sons.

Ness, I. (2016). *Southern insurgency: The coming of the global working class*. Pluto Press.

Ninan, S. (2009). *Headlines from the Heartland: Reinventing the Hindi Public Sphere*. Sage Publications.

Nobel, D. F. (2011). *Forces of production: A short history of industrial production*. Transaction Publishers.

Nussbaum, M. (2000). *Women and human development: The capabilities approach*. Cambridge University Press.

O'Neil, C. (2016). *Weapons of math destruction: How big data increases inequality and threatens democracy*. Crown.

Olson, M. (1965). *The logic of collective action: Public goods and the theory of groups*. Harvard University Press.

Ostrom, E. (1990). *Governing the Commons*. John Wiley Sons.

Pask, G. (1976). *Conversation theory: Applications in education and epistemology*. Elsevier.

Pavarala, V., & Malik, K. K. (2007). *Other voices: The struggle for community radio in India*. Sage Publications.

Piketty, T. (2014). *Capital in the twenty-first century*. Harvard Business School Press.

Piketty, T. (2020). *Capital and ideology*. Harvard University Press.

Popper, K. (1945). *Open society and its enemies*. Routledge.

Quinn, M. J. (2014). *Ethics for the information age*. Pearson.

Riley, D. (2008). *Engineering and social justice*. Morgan and Claypool Publishers.

Rodriguez, C. (2011). *Citizens' media against armed conflict: Disrupting violence in Colombia*. University of Minnesota Press.

Rokeach, M. (1973). *The nature of human values*. Free Press.

Rosling, H., Rosling, O., & Rönnlund, A. R. (2018). *Factfulness: Ten reasons we're wrong about the world – And why things are better than you think*. Sceptre.

Salter, L. (2007). *Conflicting forms of use: The potential of and limits to the use of the Internet as a public sphere*. VDM Verlag Dr. Muller and Co. KG.

Sandel, M. (2009). *Justice: What's the right thing to do*. Penguin.

Sandel, M. J. (2020). *Tyranny of merit: What's become of the Common Good*. Farrar, Straus and Giroux.

Schiffrin, A. (2017). In A. Schiffrin (Ed.), *In the service of power: Media capture and the threat to democracy*. National Endowment for Democracy.

Schumacher, E. F. (1973). *Small is beautiful: A study of economics as if people mattered*. Blond & Briggs.

Scott, J. C. (1998). *Seeing like a state: How certain schemes to improve the human condition have failed*. Yale University Press.

Scott, J. C. (2012). *Two cheers for anarchism: Six easy pieces on autonomy, dignity, and meaningful work and play*. Princeton University Press.

Sen, A. (2000). *Development as freedom*. Oxford University Press.

Sen, A. (2009). *The idea of justice*. Harvard University Press.

Sen, A., & Drèze, J. (2013). *An uncertain glory: India and its contradictions*. Allen Lane.

Smith, M. R., & Marx, L. (1996). In M. R. Smith & L. Marx (Eds.), *Does technology drive history? The dilemma of technological determinism*. The MIT Press.

de Soto, H. (2001). *The mystery of capital: Why capitalism triumphs in the West and fails everywhere else*. Black Swan.

Stiglitz, J. E. (2012). *The price of inequality: How today's divided society endangers our future*. WW Norton Company.

Sunstein, C. (2018). *#Republic: Divided democracy in the age of social media*. Princeton University Press.

Thaler, R. H., & Sunstein, C. R. (2008). *Nudge: Improving decisions about health, wealth and happiness*. Yale University Press.

Tilly, C. (1999). *Durable inequality*. University of California Press.

Toyama, K. (2015). *Geek Heresy: Rescuing social change from the cult of technology*. Perseus Books Group.

Unwin, T. (2017). *Reclaiming information and communication technologies for development*. Oxford University Press.

Vivek, S. (2014). *Delivering public services effectively: Tamil Nadu and beyond*. Oxford University Press.

Webster, F. (1995). *Theories of the information society*. Routledge.

Weizenbaum, J. (1976). *Computer power and human reason: From judgment to calculation*. W.H. Freeman and Company.

Wiener, N. (1950). *The human use of human beings: Cybernetics and society*. Houghton Mifflin.

Young, I. M. (2011). *Justice and the politics of difference*. Princeton University Press.

Zuboff, S. (2018). *The age of surveillance capitalism: The fight for a human future at the new frontier of power*. Profile Books.

Articles

Abebe, R., Barocas, S., Kleinberg, J., Levy, K., Raghavan, M., & Robinson, D. (2020). Roles for computing in social change. *Proceedings of the 2020 ACM Conference on Fairness, Accountability, and Transparency*.

Acemoglu, D. (2020). Antitrust alone won't fix the innovation problem. *Project Syndicate*. https://www.project-syndicate.org/commentary/google-antitrust-big-tech-hurdle-to-innovation-by-daron-acemoglu-2020-10

ACM. (2017). ACM statement on algorithmic transparency and accountability. *ACM*. http://www.acm.org/binaries/content/assets/public-policy/2017_joint_statement_algorithms.pdf

ACM. (2020). ACM code of ethics and professional conduct. *ACM*. https://www.acm.org/code-of-ethics

Acosta, A. M., & Pettit, J. (2014). Practice guide: A combined approach to political economy and power analysis. *SIDA*. https://www.ids.ac.uk/publications/practice-guide-a-combined-approach-to-political-economy-and-power-analysis/

ACT4D. (2019). Giant Economy Monitor. *Appropriate Computing Technologies for Development*. http://act4d.iitd.ac.in/act4dgem/index.htm

Ada Lovelace Institute and AI Council. (2021). Exploring legal mechanisms for data stewardship. *Ada Lovelace Institute and UK AI Council*. https://www.adalovelaceinstitute.org/report/legal-mechanisms-data-stewardship/

Adamic, L. A., & Glance, N. (2005). The political blogosphere and the 2004 U.S. election: Divided they blog. *Proceedings of the 3rd International Workshop on Link discovery, LinkKDD*.

Adhikari, A., Narayanan, R., Dhorajiwala, S., & Mundoli,, S. (2020). 21 Days and counting: COVID-19 lockdown, migrant workers, and the inadequacy of welfare measures in India. *Stranded Workers Action Network*. http://strandedworkers.in/mdocs-posts/21-days-and-counting-2/

Adler, P. S. (1990). Marx, machines, and skill. *Technology and Culture*, *31*(4), 780–812.

Agarwal, B. (2018). Can group farms outperform individual family farms? Empirical insights from India. *World Development*, *108*, 57–73.

Agarwal, D. (2017). Sedition and social media: Section 124A needs a relook as it could be a tool for harassment. *Firstpost*. https://www.firstpost.com/india/sedition-and-social-media-section-124a-needs-a-relook-as-it-could-be-a-tool-for-harassment-3945485.html

Agrawal, S., Banerjee, S., & Sharma, S. (2017). Privacy and security of Aadhaar: A computer science perspective. *Economic and Political Weekly*, *52*(37).

Agre, P. E. (1997). Toward a critical technical practice: Lessons learned in trying to reform AI. In G. Bowker, L. Gasser, L. Star, & B. Turner (Eds.), *Bridging the great divide: Social science, technical systems, and cooperative work*. Erlbaum.

AgStack. (2021). AgStack: A Linux Foundation Project. https://agstack.org/

AIID. (2020). AI incident database. https://incidentdatabase.ai/

Aiken, E., Bellue, S., Karlan, D., Udry, C. R., & Blumenstock, J. (2021). Machine learning and mobile phone data can improve the targeting of humanitarian assistance. *NBER Working Paper No. 29070. National Bureau of Economic Research*.

Aithala, V. (2020). Idea of social stock exchanges for India. *The India Forum*. https://www.theindiaforum.in/article/social-stock-exchanges-india-sebi-s-promise

Ali, A. T. (2021). The Indian farmers' movement has shattered Narendra Modi's strongman image. *Jacobin*. https://jacobinmag.com/2021/11/indian-farmers-movement-neoliberal-farm-bills-modi-bjp

Allen, C., Wallach, W., & Smit, I. (2006). Why machine ethics? *IEEE Intelligent Systems, 21*(4).

Allmer, T. (2011). A critical contribution to theoretical foundations of privacy studies. *Journal of Information, Communication Ethics in Society, 9*(12), 83–101.

Alphabet Workers Union. (2021). Alphabet Workers Union: Our mission. https://alphabetworkersunion.org/principles/mission-statement/

Angwin, J., Larson, J., Mattu, S., & Kirchner, L. (2016). Machine bias. *ProPublica*. https://www.propublica.org/article/machine-bias-risk-assessments-in-criminal-sentencing

Appadurai, A. (2020). Goodbye citizenship, hello 'statizenship'. *The Wire*. https://thewire.in/rights/goodbye-citizenship-hello-statizenship

Arndt, C. (2018). White-collar unions and attitudes towards income inequality, redistribution, and state–market relations. *European Sociological Review, 34*(6), 675–693.

Arnstein, S. (1969). A ladder of citizen participation. *Journal of the American Planning Association, 35*(4), 216–224.

Arsenault, A. H., & Castells, M. (2008). The structure and dynamics of global multi-media business networks. *International Journal of Communication, 2*, 707–728.

Arun, C. (2019). On WhatsApp, rumours, and lynchings. *Economic Political Weekly, 54*(6).

Asthana, S., Kumar, R., Bhagwan, R., Bansal, C., Bird, C., Maddila, C., Mehta, S., & Ashok, B. (2019). WhoDo: Automating reviewer suggestions at scale. *Proceedings of the 2019 27th ACM Joint Meeting on European Software Engineering Conference and Symposium on the Foundations of Software Engineering, ESEC/FSE.*

Athey, S., Catalini, C., & Tucker, C. (2017). The digital privacy paradox: Small money, small costs, small talk. *NBER Working Paper No. 23488. National Bureau of Economic Research.*

Auston, L. M. (2019). Re-reading Westin. *Theoretical Inquiries, 20*, 53–81.

Avelino, F., Wittmayer, J. M., Pel, B., Weaver, P., Dumitru, A., Haxeltine, A., Kemp, R., Jørgensen, M. S., Bauler, T., Ruijsink, S., & O'Riordan, T. (2017). Transformative social innovation and (dis)empowerment. *Technological Forecasting and Social Change, 145*, 195–206.

Avle, S., Lindtner, S., & Willians, K. (2017). How methods make designers. *Proceedings of the 2017 ACM Conference on Human Factors in Computing Systems, ACM CHI.*

Bakshy, E., Messing, S., & Adamic, L. (2015). Exposure to ideologically diverse news and opinion on Facebook. *Science, 348*(6239), 1130–1132.

Bakshy, E., Rosenn, I., Marlow, C., & Adamic, L. (2012). The role of social networks in information diffusion. *Proceedings of the 21st International Conference on World Wide Web.*

Balabanian, N. (2000). Controlling technology: Should we rely on the marketplace? *IEEE Technology and Society Magazine, 19*(2).

Banerjee, A., Karlan, D., & Zinman, J. (2015). Six randomized evaluations of microcredit: Introduction and further steps. *American Economic Journal: Applied Economics, 7*(1).

Banerjee, S., & Sagar, A. (2021). What we must consider before digitising India's healthcare. *The Indian Express*. https://indianexpress.com/article/opinion/columns/national-digital-health-mission-harsh-vardhan-healthcare-sector-7218715/

Bansal, C., Singla, A., Singh, A. K., Ahlawat, H. O., Jain, M., Singh, P., Kumar, P., Saha, R., Taparia, S., & Seth, A. (2020). Characterizing the evolution of Indian cities using satellite imagery and open street maps. *Proceedings of the 3rd ACM SIGCAS Conference on Computing and Sustainable Societies, ACM COMPASS.*

Bansal, C., Ahlawat, H. O., Jain, M., Prakash, O., Mehta, S. A., Singh, D., Baheti, H., Singh, S., & Seth, A. (2021). IndiaSat: A pixel-level dataset for land-cover classification on three satellite systems – Landsat-7, Landsat-8, and Sentinel-2. *Proceedings of the 4th ACM SIGCAS Conference on Computing and Sustainable Societies, ACM COMPASS.*

Bansal, C., Jain, A., Barwaria, P., Choudhary, A., Singh, A., Gupta, A., & Seth, A. (2020). Temporal prediction of socio-economic indicators using satellite imagery. *Proceedings of the 7th ACM IKDD CoDS and 25th COMAD Conference.*

Barak, M. E. M. (2020). The practice and science of social good: Emerging paths to positive social impact. *Research on Social Work Practice*, *30*(2), 139–150.

Barboni, G., Field, E., Pande, R., Rigol, N., Schaner, S., & Troyer, C. (2018). A tough call: Understanding barriers to and impacts of women's mobile phone adoption in India. *Report. Harvard Kennedy School.*

Barocas, S., & Nissenbaum, H. (2014). Computing ethics: Big Data's end run around procedural privacy protections. *Communications of the ACM*, *57*(11).

Batliwala, S. (2019). All about power: Understanding social power and power structures. *CREA*. https://reconference.creaworld.org/wp-content/uploads/2019/05/All-About-Power-Srilatha-Batliwala.pdf

Bauwens, M., & Kostakis, V. (2014). From the communism of capital to capital for the commons: Towards an open co-operativism. *tripleC: Communication, Capitalism, and Critique*, *12*(1).

Bauwens, M., & Niaros, V. (2017). Value in the commons economy: Developments in open and contributory value accounting. *Heinrich Böll Stiftung*. https://www.boell.de/en/2017/02/01/value-commons-economy-developments-open-and-contributory-value-accounting

Bengtsson, F., & Lundstrom, J. E. (2013). ANT-Maps: Visualizing perspectives of business and information systems. *International Conference on Information Systems.*

Benjamin, S., Bhuvaneswari, R., Rajan, P., & Manjunatha. (2007). Bhoomi: 'E-Governance', or, an anti-politics machine necessary to globalize Bangalore? *CASUM-m Working Paper. Collaborative for the Advancement of Studies in Urbanism through Mixed Media*. https://casumm.files.wordpress.com/2008/09/bhoomi-e-governance.pdf

Bennett, W. L., & Iyengar, S. (2008). A new era of minimal effects? The changing foundations of political communication. *Journal of Communication*, *58*, 707–731.

Berdichevsky, D., & Neuenschwander, E. (1999). Toward an ethics of persuasive technology. *Communications of the ACM*, *42*(5).

Berg, P. (2008). Asilomar 1975: DNA modification secured. *Nature*, *455*, 290–291.

Best, M., & Kumar, R. (2008). Sustainability failures of rural telecenters: Challenges from the Sustainable Access in Rural India (SARI) Project. *Information Technologies and International Development*, *4*(4), 31–45.

Betterplace. (2020). Betterplace. https://www.betterplace.co.in/

Bird, S., Hutchinson, B., Kenthapadi, K., Kiciman, E., & Mitchell, M. (2019). Fairness-aware machine learning in practice (tutorial). *ACM SIGKDD International Conference on Knowledge Discovery and Data Mining.*

Birkinbine, B. J. (2018). Commons praxis: Towards a critical political economy of the digital commons. *tripleC: Communication, Capitalism, and Critique*, *16*(1), 290–305.

Bjerknes, G., & Bratteteig, T. (1995). User participation and democracy: A discussion of Scandinavian research on system development. *Scandinavian Journal of Information Systems*, *7*(1), 73–98.

Blomberg, J., Giacomi, J., Mosher, A., & Swenton-Wall, P. (1993). Ethnographic field methods and their relation to design. In D. Schuler & A. Namioka (Eds.), *Participatory design: Principles and practices*. Lawrence Erlbaum Associates.

Bodker, S., & Kyng, M. (2018). Participatory design that matters: Facing the big issues. *ACM Transactions on Computer–Human Interaction*, *25*(1), 1–31.

Bond, R. M., Fariss, C. J., Jones, J. J., Kramer, A. D. I., Marlow, C. J., Settle, E., & Fowler, J. H. (2012). A 61-million-person experiment in social influence and political mobilization. *Nature*, *489*, 295–298.

Bosc, P. (2018). Empowering through collective action. *International Fund for Agriculture Development (IFAD) Research Series*. https://www.ifad.org/documents/38714170/40797323/29_Research_web.pdf

Bosch, T. (2010). Theorizing citizens' media: A rhizomatic approach. In C. Rodriguez, D. Kidd, & L. Stein (Eds.), *Making our media: Global initiatives toward a democratic public sphere*. Hamptom Press.

Braa, J., Monteiro, E., & Sahay, S. (2004). Networks of action: Sustainable health information systems across developing countries. *MIS Quarterly*, *28*(3), 337–362.

Bryan, L. L. (2007). The new metrics of corporate performance: Profit per employee. *McKinsey Quarterly*. https://www.mckinsey.com/business-functions/strategy-and-corporate-finance/our-insights/the-new-metrics-of-corporate-performance-profit-per-employee

Buchanan, L., & Seshagiri, A. (2016). How Uber uses psychological tricks to push its drivers' buttons. *New York Times*. https://www.nytimes.com/interactive/2017/04/02/technology/uber-drivers-psychological-tricks.html

Bullock, J. B. (2019). Artificial intelligence, discretion and bureaucracy. *American Review of Public Administration*, *1*, 751–761.

Buolamwini, J., & Gebru, T. (2018). Gender shades: Intersectional accuracy disparities in commercial gender classification. *Machine Learning Research*, *85*, 1–15.

Butler, B., Sproull, L., Kiesler, S., & Kraut, R. (2007). Community effort in online groups: Who does the work and why? In S. Weisband & L. Atwater (Eds.), *Leadership at a distance*. Routledge.

Bynum, T. W. (2000). A very short history of computer ethics. *Newsletter on Philosophy and Computing, American Philosophical Association*.

Cadwalladr, C. (2018). The Cambridge analytica files. *The Guardian*. https://www.theguardian.com/news/series/cambridge-analytica-files

Cairns, P., & Thimbleby, H. (2003). The diversity and ethics of HCI. *Transactions of Computer–Human Interactions*.

Callon, M. (1986). Some elements of a sociology of translation: Domestication of the Scallops and the fishermen of St Brieuc Bay. In J. Law (Ed.), *Power, action and belief: A new sociology of knowledge?* Routledge.

Campbell, D. T. (1991). Methods for the experimenting society. *Evaluation Practice*, *12*(3), 223–260.

Cantillon, E. (2017). Broadening the market design approach to school choice. *Oxford Review of Economic Policy*, *33*(4), 613–634.

Carpentier, N., Lie, R., & Servaes, J. (2003). Community media: Muting the democratic media discourse? *Continumm: Journal of Media and Communication Studies*, *17*(1), 51–68.

Carpini, M. X. D., Cook, F. L., & Jacobs, L. R. (2004). Public deliberation, discursive participation, and citizen engagement: A review of the empirical literature. *Annual Review of Political Science*, *7*, 315–344.

Carswell, G., Chambers, T., & Neve, G. D. (2019). Waiting for the state: Gender, citizenship and everyday encounters with bureaucracy in India. *Politics and Space*, *37*(4), 597–616.

Castells, M. (2016). A sociology of power: My intellectual journey. *Annual Review of Sociology*, *42*, 1–19.

Centivany, A., & Glushko, B. (2015). "Popcorn tastes good": Participatory policymaking and Reddit's "AMAgeddon". *Proceedings of the 2015 ACM Conference on Human Factors in Computing Systems, ACM CHI*.

Chakrabarti, S. (2014). How structure shapes content, or why the 'Hindi turn' of Star Plus became the 'Hindu' turn. *Media, Culture and Society, 34*(4).

Chakraborty, D. (2018). Building ICT based information flows to improve citizen–government engagement. *Ph.D. thesis, Indian Institute of Technology Delhi.*

Chakraborty, D., Ahmad, M. S., & Seth, A. (2017). Findings from a civil society mediated and technology assisted grievance redressal model in rural India. *Proceedings of the 2017 International Conference on Information and Communication Technologies and Development (ICTD).*

Chakraborty, D., Gupta, A., Gram Vaani Team, & Seth, A. (2019). Experiences from a mobile-based behaviour change campaign on maternal and child nutrition in rural India. *Proceedings of the 2019 International Conference on Information and Communication Technologies and Development (ICTD).*

Chakraborty, S., Pal, J., Chandra, P., & Romero, D. M. (2018). Political Tweets and Mainstream News impact in India: A mixed methods investigation into political outreach. *Proceedings of the 1st ACM SIGCAS Conference on Computing and Sustainable Societies, ACM COMPASS.*

Chandrasekharan, E., Pavalanathan, U., Srinivasan, A., Glynn, A., Eisenstein, J., & Gilbert, E. (2018). You can't stay here: The efficacy of Reddit's 2015 Ban examined through hate speech. *Proceedings of the ACM Conference On Computer-Supported Cooperative Work And Social Computing, CSCW.*

Chandrasekharan, E., Samory, M., Jhaver, S., Charvat, H., Bruckman, A., Lampe, C., Eisenstein, J., & Gilbert, E. (2018). The Internet's hidden rules: An empirical study of Reddit norm violations at micro, meso, and macro scales. *Proceedings of the ACM Conference On Computer-Supported Cooperative Work And Social Computing, CSCW.*

Chaudhuri, D., & Ghosh, P. (2020). Old wine in a new bottle? A look at how the evolution of organic farming in India is harming farmers by 'deskilling' them. *India Development Review*. https://idronline.org/problem-withorganic-farming/

Chell, E., Spence, L. J., Perrini, F., & Harris, J. D. (2016). Social entrepreneurship and business ethics: Does social equal ethical? *Journal of Business Ethics, 133.*

Chenoweth, E. (2016). How social media helps dictators. *Foreign Policy*. https://foreignpolicy.com/2016/11/16/how-social-media-helps-dictators/

Cho, J., Ahmed, S., Keum, H., Choi, Y. J., & Lee, J. H. (2018). Influencing myself: Self-reinforcement through online political expression. *Communication Research, 45*, 83–111.

Chouldechova, A., & Roth, A. (2018). The Frontiers of fairness in machine learning. *Computing Community Consortium Workshop*. https://arxiv.org/abs/1810.08810

Christiansen, E. (2014). From "ethics of the eye" to "ethics of the hand" by collaborative prototyping. *Journal of Information Communication and Ethics in Society, 10*(1).

Cinelli, M., Morales, G. D. F., Galeazzi, A., Quattrociocchi, W., & Starnini, M. The echo chamber effect on social media. *PNAS* 118.9

Clark, H., & Taplin, D. (2012). Theory of change basics: A primer on theory of change. *ActKnowledge*. https://www.theoryofchange.org/wp-content/uploads/toco_library/pdf/ToCBasics.pdf

Clayton, J. (2021). Google threatens to withdraw search engine from Australia. *BBC*. https://www.bbc.com/news/world-australia-55760673

Cohen, J. E. (2013). What privacy is for. *Harvard Law Review, 1904. 126*, 1904–1933

Conger, K. (2021). Hundreds of Google employees unionize, culminating years of activism. *New York Times*. https://www.nytimes.com/2021/01/04/technology/google-employees-union.html

Conover, M. D., Ratkiewicz, J., Francisco, M., Goncalves, B., Flammini, A., & Menczer, F. (2011). Political polarization on Twitter. *Proceedings of the International AAAI Conference on Web and Social Media, ICWSM.*

Crawford, K. (2016). Can an algorithm be agonistic? Ten scenes from life in calculated publics. *Science, Technology and Human Values, 4*(1), 77–92.

Crawford, K., & Boyd, D. (2012). Critical questions for Big Data. *Information, Communication and Society*, *15*(5).

Crawford, K., & Gillespie, T. (2016). What is a flag for? Social media reporting tools and the vocabulary of complaint. *New Media and Society*, *18*(3), 410–428.

Cross, T., Gupta, N., Liu, B., Nair, V., Kumar, A., Kuttan, R., Ivatury, P., Chen, A., Lakshman, K., Rodrigues, R., D'Souza, G., Chittamuru, D., Rao, R., Rade, K., Vadera, B., Shah, D., Choudhary, V., Chadha, V., Shah, A., Kumta, S., Dewan, P., Thomas, B., & Thies, W. (2019). 99DOTS: A low-cost approach to monitoring and improving medication adherence. *Proceedings of the 2019 International Conference on Information and Communication Technologies and Development (ICTD)*.

Dafermos, G. (2016). Digital commons: Cyber-commoners, peer producers and the project of a post-capitalist transition. *Heteropolitics: Refiguring the Common and the Political*. https://heteropolitics.net/index.php/2020/12/31/cyber-commoners-peer-producers-and-the-project-of-a-post-capitalist-transition/

Dahlbom, B., & Mathiassen, L. (1997). The future of our profession. *Communications of the ACM*, *40*(6). https://doi.org/10.1145/255656.255706

Dantec, C. A., Poole, E. S., & Wyche, S. P. (2009). Values as lived experience: Evolving value sensitive design in support of value discovery. *Proceedings of the 2009 ACM Conference on Human Factors in Computing Systems, ACM CHI*.

Dark Patterns. (2020). Dark Patterns. https://www.darkpatterns.org/

Dattani, K. (2019). "Governtrepreneurism" for good governance: The case of Aadhaar and the India Stack. *Royal Geographical Society*, *52*(2), 411–419.

De, R., Pal. A., Sethi, R., Reddy, S. K., & Chitre, C. (2018). ICT4D research: A call for a strong critical approach. *Information Technology for Development*, *24*(1), 63–94.

DFID. (2009). Political economy analysis: How to note. *A DFID Practice Paper*. https://www.odi.org/sites/odi.org.uk/files/odi-assets/events-documents/3797.pdf

Dharmakumar, R. (2017). For whom does the India Stack bell toll? *The Ken*. https://the-ken.com/story/india-stack-bell-toll/

Dhillon, A. (2018). 'My life is spent in this car': Uber drives its Indian workers to despair. *The Guardian*. https://www.theguardian.com/global-development/2018/dec/04/my-life-is-spent-in-this-car-uber-drives-indian-workers-to-despair

Diakopoulos, N. (2017). Algorithmic accountability reporting: On the investigation of black boxes. *Tow Center for Digital Journalism, Columbia University*.

van Dijk, T. A. (1989). Structures of discourse and structures of power. *Annals of the International Communication Association*, *12*(1), 18–59.

Doward, J. (2018). The Big Tech Backlash. *The Guardian*. https://www.theguardian.com/technology/2018/jan/28/tech-backlash-facebook-google-fake-news-business-monopoly-regulation

Dreher, T. (2009). Listening across difference: Media and multiculturalism beyond the politics of voice. *Continumm: Journal of Media and Cultural Studies*, *23*(4), 445–458.

Drèze, J. (2021). There is an urgent need for safeguards against unfair discontinuation of social benefits. *The Indian Express*. https://indianexpress.com/article/opinion/columns/aadhaar-linking-public-welfare-schemes-pds-system-7280621/

Drèze, J., & Khera, R. (2015). Understanding leakages in the public distribution system. *Economic and Political Weekly*, *50*(7).

Duboc, L., McCord, C., Becker, C., & Ahmed, S. I. (2020). Critical requirements engineering in practice. *IEEE Software*, *37*(1), 17–24.

Duquenoy, P., & Thimbleby, H. (1999). Justice and design. In M. A. Sasse & C. Johnson (Eds.), *Proceedings of human–computer interaction* (pp. 281–286).

Economist. (2020). Starting over again: The COVID-19 pandemic is forcing a rethink in macroeconomics. https://www.economist.com/briefing/2020/07/25/the-covid-19-pandemic-is-forcing-a-rethink-in-macroeconomics

Edelman. (2020). Edelman Trust Barometer: 2020. *20th Annual Edelman Trust Barometer*. https://www.edelman.com/sites/g/files/aatuss191/files/2020-01/2020%5C%20Edelman%5C%20Trust%5C%20Barometer%5C%20Global%5C%20Report_LIVE.pdf

Edmund, D. S. (2015). Phishing in Jamtara: What does it take to carry out online fraud? *The Indian Express*. https://indianexpress.com/article/india/india-news-india/phishing-in-jamtara-what-does-it-take-to-carry-out-online-fraud/

Ekbia, H., & Nardi, B. (2015). The political economy of computing: The elephant in the HCI room. *ACM Interactions, 22*(6).

Elstub, S. (2006). A double-edged sword: The increasing diversity of deliberative democracy. *Contemporary Politics, 12*, 301–319.

Epstein, R., & Robertson, R. E. (2015). The search engine manipulation effect (SEME) and its possible impact on the outcomes of elections. *PNAS, 112*(33), 4512–4521.

Eslami, M., Kumaran, S. R. K., Sandvig, C., & Karahalios, K. (2018). Communicating algorithmic process in online behavioral advertising. *Proceedings of the 2018 ACM Conference on Human Factors in Computing Systems, ACM CHI.*

Espejo, R. (2014). Cybernetic argument for democratic governance: Cybersyn and Cyberfolk. *Social Systems and Design.*

Ethical Design Toolkit. (2020). Ethical design toolkit. https://www.ethicsfordesigners.com

Farcane, N., Deliu, D., & Bureana, E. (2019). A corporate case study: The application of Rokeach's value system to corporate social responsibility (CSR). *Sustainability, 11.*

Farooq, U., Merkel, C. B., Nash, H., Rosson, M. B., Carroll, J. M., & Xiao, L. (2005). Participatory design as apprenticeship: Sustainable watershed management as a community computing application. *Hawaii international conference on system sciences.*

FATML. (2019). Fairness, accountability, and transparency in machine learning. https://www.fatml.org/

Fleming, A., Jakku, E., Lim-Camacho, L., Taylor, B., & Thorburn, P. (2018). Is Big-Data for big farming or for everyone? Perceptions in the Australian grains industry. *Agronomy for Sustainable Development, 38* (24).

Floridi, L. (1999). Information ethics: On the philosophical foundation of computer ethics. *Ethics and Information Technology, 1*, 33–52.

Floridi, L. (2018). AI as a force for good. *World Summit for AI*. https://www.youtube.com/watch?v=_Fbvq-HeMNo

Flyverbom, M., Christensen, L. T., & Hansen, H. K. (2015). The transparency–power nexus: Observational and regularizing control. *Management Communication Quarterly, 29*(3), 385–410.

Fogg, B. J. (1973). A behavior model for persuasive design. *Proceedings of the 1998 ACM Conference on Human Factors in Computing Systems, ACM CHI.*

Fogg, B. J. (1988). The elements of computer credibility. *Proceedings of the 2018 ACM Conference on Human Factors in Computing Systems, ACM CHI.*

Fogg, B. J. (2003). Computers as persuasive social actors. *Persuasive Technology.*

Følstad, A., Brandtzaeg, P. B., Feltwell, T., Law, E. L.-C., Tscheligi, M., & Luger, E. A. (2018). Chatbots for social good. *Proceedings of the 2018 ACM Conference on Human Factors in Computing Systems, ACM CHI.*

Fox, J. (2018). Why German corporate boards include workers. *Bloomberg*. http://www.bloomberg.com/view/articles/2018-08-24/why-german-corporate-boards-include-workers-for-co-determination

Frauenberger, C., Rauhala, M., & Fitzpatrick, G. (2017). In-action ethics. *Interacting with Computers, 29*(2), 220–236.

Freuler, J. O. (2018). Techlash: Why Facebook's approach to FakeNews ultimately fails. *Open Democracy*. https://www.opendemocracy.net/en/digitaliberties/techlash-why-facebook-s-approach-to-fakenews-ultimately-fails/

Friedman, B., Kahn, P. H., & Borning, A. (2013). Value sensitive design and information systems. In P. Zhang & D. Gatella (Eds.), *Human computer interaction in management information systems: Foundations*. M.E. Sharpe.

Fuchs, C. (2020). Towards a critical theory of communication as renewal and update of Marxist humanism in the age of digital capitalism. *Journal for Theory of Social Behavior*, *50*, 335–356

Fuchs, C. (2021a). Cornel West and Marxist humanism. *Critical Sociology*, *47*, 1219–1243

Fuchs, C. (2021b). The digital commons and the digital public sphere: How to advance digital democracy today. *Westminster Papers in Communication and Culture*, *16*(1), 9–26.

Fuchs, C., & Sandoval, M. (2015). The political economy of capitalist and alternative social media. In C. Atton (Ed.), *Routledge companion to alternative and community media* (pp. 165–175). Routledge.

Funke, M., Schularick, M., & Trebesch, C. (2021). The cost of populism: Evidence from history. *VoxEU*. https://voxeu.org/article/cost-populism-evidence-history

Gabor, D., & Brooks, S. (2016). The digital revolution in financial inclusion: International development in the Fintech era. *New Political Economy*, *22*(4).

Galic, M., Timan, T., & Koops, B. (2017). Bentham, Deleuze and beyond: An overview of surveillance theories from the panopticon to participation. *Philosophical Transactions*, *30*, 9–37.

Gansen, K. V., Valayer, C., & Allessie, D. (2018). Digital platforms for public services. *ISA Programme, European Union*. https://bit.ly/2RHf8H0

Gaventa, J. (2019). Applying power analysis: Using the 'Powercube' to explore forms, levels and spaces. *The Changing Faces of Power 1979–2019*.

GDPR. (2016). General data protection regulation. https://gdpr-info.eu/

Geiger, R. S. (2015). Does Facebook have civil servants? On governmentality and computational social science. *CSCW workshop on Ethics for Studying Sociotechnical Systems in a Big Data World*.

Gelb, A., & Mukherjee, A. (2021). A COVID vaccine certificate: Building on lessons from digital ID for the digital yellow card. *Center for Global Development*. https://www.cgdev.org/publication/covid-vaccine-certificate-building-lessons-digital-id-digital-yellow-card

Gillespie, T. (2010). The politics of 'platforms'. *New Media and Society*, *12*(3), 347–364.

Gillespie, T. (2013). The relevance of algorithms. In T. Gillespie, P. J. Boczkowski, & K. A. Foot (Eds.), *Media technologies: Essays on Communication, Materiality, and Society, MIT Press*.

Gillespie, T. (2014). Algorithm [draft] digitalkeywords. *Culture Digitally*. https://culture-digitally.org/2014/06/algorithm-draft-digitalkeyword/

Gilpin, L. H., Bau, D., Yuan, B. Z., Bajwa, A., Specter, M., & Kagal, L. (2018). Explaining explanations: An overview of interpretability of machine learning. *Proceedings of the 2018 IEEE International Conference on Data Science and Advanced Analytics, DSAA*.

Gisler, P., & Kurath, M. (2011). Paradise lost? "Science" and "the public" after Asilomar. *Science, Technology, and Human Values*, *36*(2), 213–243.

GoI. (2020a). Consultation paper on IDEA: India Digital Ecosystem of Agriculture. *Department of Agriculture, Government of India*. https://agricoop.nic.in/sites/default/files/IDEA%5C%20Concept%5C%20Paper_mod31052021_2.pdf

GoI. (2020b). National Digital Health Blueprint. *Ministry of Health and Family Welfare, Government of India*. https://main.mohfw.gov.in/newshighlights/final-report-national-digital-health-blueprint-ndhb

GoI. (2020c). Strategy for National Open Digital Ecosystems (NODE): Consultation paper. *Ministry of Electronics and Information Technology, Government of India*. https://www.medianama.com/wp-content/uploads/mygov_1582193114515532211.pdf

Goldsmith, J., & Burton, E. (2017). Why teaching ethics to AI practitioners is important. *AAAI Workshop on AI, Ethics, and Society*.

Gonzalez, C. (2015). Environmental justice, human rights, and the Global South. *Santa Clara Journal of International Law*, *151*.

Goodman, B., & Flaxman, S. (2017). European union regulations on algorithmic decision-making and a 'right to explanation'. *AI Magazine*, *38*(5), 50–57.

Gopalaswamy, A. K., Babu, M. S., & Dash, U. (2016). Systematic review of quantitative evidence on the impact of microfinance on the poor in South Asia. *EPPI-Centre, Social Science Research Unit, University College London*. http://eppi.ioe.ac.uk/CMS/Portals/0/PDF%5C%20reviews%5C%20and%5C%20summaries/Microfiannce-ST%5C%20-%5C%20IITM2.pdf

Gotterbarn, D., & Rogerson, S. (2005). Responsible risk assessment with software development: Creating the software development impact statement. *Communication of the Association for Information Systems, 15*, 730–750.

GRAIN. (2020). Digital fences: The financial enclosure of farmlands in South America. https://grain.org/en/article/6529-digital-fences-the-financial-enclosure-of-farmlands-in-south-america

Gram Vaani. (2017). Integrated model for grievance-redressal and access to justice. *UNDP Final Project Report* available upon request.

Gram Vaani. (2019). The Mobile Vaani Manifesto. *Gram Vaani*. https://www.cse.iitd.ac.in/~aseth/MV-manifesto-2019.pdf

Granovetter, M. (1973). The strength of weak ties. *Annual Journal of Sociology, 78*(6), 1360–1380.

Green, B. (2020). Data science as political action: Grounding data science in a politics of justice. *Journal of Social Computing*. Available at SSRN https://dx.doi.org/10.2139/ssrn.3658431

Green, B., & Chen, Y. (2019). Disparate interactions: An algorithm-in-the-loop analysis of fairness in risk assessments. *Proceedings of the 2019 ACM Conference on Fairness, Accountability, and Transparency*.

Greene, D., Hoffmann, A. L., & Stark, L. (2019). Better, nicer, clearer, fairer: A critical assessment of the movement for ethical artificial intelligence and machine learning. *Hawaii International Conference on System Sciences*.

Gregory, J. (2003). Scandinavian approaches to participatory design. *International Journal of Engineering Education, 19*(1).

Grevet, C., Terveen, L., & Gilbert, E. (2014). Managing political differences in social media. *Proceedings of the ACM Conference On Computer-Supported Cooperative Work And Social Computing, CSCW*.

Grimmelmann, J. (2015). Law and ethics of experiments on social media users. *Colorado Technology Law Journal, 219*.

Grosser, B. (2014). What do metrics want? How quantification prescribes social interaction on Facebook. *Computational Culture, 4*.

Gupta, A., Narayanan, A., Bhutani, A., Seth, A., Johri, M., Kumar, N., Ahmad, S., Rahman, M., Enoch, L., Kumar, A., Sharma, D., Kumar, A., Sharma, A., & Pappu, L. R. (2021). Delivery of social welfare entitlements in India: Unpacking exclusion, grievance redress, and the role of civil society organisations. *Azim Premji University COVID-19 Research Funding Programme 2020*. https://gramvaani.org/?p=3919

Haider, A., & Mohandesi, S. (2013). Workers' inquiry: A genealogy. *Viewpoint Magazine*. https://viewpointmag.com/2013/09/27/workers-inquiry-a-genealogy/

Halfaker, A., Geiger, R. S., & Terveen, L. (2014). Snuggle: Designing for efficient socialization and ideological critique. *Proceedings of the 2014 ACM Conference on Human Factors in Computing Systems, ACM CHI*.

Hanel, P. H. P., Litzellachner, L. F., & Gregory, M. R. (2018). An empirical comparison of human value models. *Frontiers in Psychology, 9*.

Hansen, N. B., Dindler, C., Halskov, K., & Iversen, O. S. (2019). How participatory design works: Mechanisms and effects. *Proceedings of the 31st Australian Conference on Human-Computer-Interaction, OZCHI*.

Harding, M., Knowles, B., Davies, N., & Rouncefield, M. (2015). HCI, civic engagement and trust. *Proceedings of the 2015 ACM Conference on Human Factors in Computing Systems, ACM CHI*.

Hardt, M. (2010). The common in communism. *Rethinking Marxism*, *22*(3).

Hardy, C., & Leiba-O'Sullivan, S. (1998). The power behind empowerment: Implications for research and practice. *Human Relations*, *51*(4), 451–483.

Harriss-White, B. (2017). Demonetisation has permanently damaged India's growth story: Barbara Harriss-White. *The Wire*. https://thewire.in/economy/demonetisation-interview-black-money

Harvey, D. (2003). The fetish of technology: Causes and consequences. *Macalester International*, *13*(7), 3–30. http://digitalcommons.macalester.edu/macintl/vol13/iss1/7

Haxeltine, A., Avelino, F., Pel, B., Dumitru, A., Kemp, R., Longhurst, N., Chilvers, J., & Wittmayer, J. M. (2016). A framework for transformative social innovation. *TRANSIT Working Paper No. 5*.

Hayes, G. (2011). The relationship of action research to human–computer interaction. *ACM Transactions on Computer–Human Interaction*, *18*(3).

Hayes, N., & Westrup, C. (2014). Consultants as intermediaries and mediators in the construction of information and communication technologies for development. *Information Technologies and International Development*, *10*(1).

Heeks, R. (1999). The tyranny of participation in information systems: Learning from development projects. *Development Informatics: Working Paper 4*.

Hindman, M., Tsioutsiouliklis, K., & Johnson, J. A. (2003). Googlearchy: How a few heavily-linked sites dominate politics on the web. *Annual meeting of the Midwest Political Science Association*.

Hirsch, T. (2016). Surreptitious communication design. *Design Issues*, *37*(2).

Hirschheim, R., & Klein, H. K. (1989). Four paradigms of information systems development. *Communications of the ACM*, *32*(10).

Hodges, D. C. (1965). Marx's contribution to Humanism. *Science and Society*, *29*(2), 173–191.

Hodson, D. (2010). Science education as a call to action. *Canadian Journal of Science, Mathematics and Technology Education*, *10*(3), 197–206.

Holbert, R. L., Garrett, R. K., & Gleason, L. S. (2010). A new era of minimal effects? A response to Bennett and Iyengar. *Journal of Communication*, *40*, 15–34.

van den Hoven, J. (2010). The use of normative theories in computer ethics. In L. Floridi (Ed.), *The Cambridge handbook of information and computer ethics* (Chapter 4).

Huff, C., & Rogerson, S. (2005). Craft and reform in moral exemplars in computing. *Proceedings of the international conference on the ethical and social impacts of ICT*.

Hunjan, R., & Pettit, J. (2011). Power: A practical guide for facilitating social change. *Carnegie United Kingdom Trust*. https://www.carnegieuktrust.org.uk/publications/power-a-practical-guide-for-facilitating-social-change/

Hunnicutt, G. (2009). Varieties of patriarchy and violence against women: Resurrecting "patriarchy" as a theoretical tool. *Violence Against Women*, *15*(5), 553–573.

Iazzolino, G., Ouma, M., & Mann, L. (2020). A digital new deal against corporate hijack of the post-Covid 19 future. *A Digital New Deal: Visions of Justice in a Post-COVID World*. https://itforchange.net/digital-new-deal/2020/10/29/a-digital-new-deal-against-corporate-hijack-of-the-post-covid-19-future/

ICDPPC. (2018). Declaration on ethics and data protection in artificial intelligence. *International conference of data protection and privacy commissioners*. https://edps.europa.eu/sites/edp/files/publication/icdppc-40th_ai-declaration_adopted_en_0.pdf

Ideo. (2008). Design kit: The human centered design toolkit. *Ideo*. https://www.ideo.com/post/design-kit

IEEE. (2014). IEEE code of conduct. *IEEE*. https://www.ieee.org/content/dam/ieee-org/ieee/web/org/about/ieee_code_of_conduct.pdf

IIF. (2020). A public brief and analysis on the Personal Data Protection Bill, 2019. *Save Our Privacy, Internet Freedom Foundation*. https://saveourprivacy.in/media/all/Brief-PDP-Bill-25.12.2020.pdf

Iivari, N., & Kuuti, K. (2017). Critical design research and information technology: Searching for empowering design. *ACM Conference on Designing Interactive Information Systems, DIS*.

Irani, L., Vertesi, J., Dourish, P., Philip, K., & Grinter, R. E. (2010). Postcolonial computing: A lens on design and development. *Proceedings of the 2010 ACM Conference on Human Factors in Computing Systems, ACM CHI*.

Irwin, J. (2015). Ethical consumerism isn't dead, it just needs better marketing. *Harvard Business Review*. https://hbr.org/2015/01/ethical-consumerism-isnt-dead-it-just-needs-better-marketing

Issac, M. (2019). Dissent erupts at Facebook over hands-off stance on political ads. *New York Times*. https://www.nytimes.com/2019/10/28/technology/facebook-mark-zuckerberg-political-ads.html

Jatav, M., & Jajoria, D. (2020). The Mathadi model and its relevance for empowering unprotected workers. *Economic and Political Weekly*, *55*(19), 155–170.

Jayaraman, P. P., Yavari, A., Georgakopoulos, D., Morshed, A., & Zaslavsky, A. (2016). Internet of Things platform for smart farming: Experiences and lessons learnt. *Sensors*, *16*.

Jian, L., & Usher, N. (2014). Crowd-funded journalism. *Journal of Computer Mediated Communication*, *19*.

Johnson, T. A. (2020). Rapid expansion, random wage cuts: Behind the Wistron violence. *The Indian Express*. https://indianexpress.com/article/explained/wistron-violence-apple-bengaluru-7117474/

Jorna, F., & Wagenaar, P. (2007). The 'iron cage' strengthened? Discretion and digital discipline. *Public Administration*, *85*(1), 189–214.

Jose, V. K. (2020). The boiled frog: Indian media and possible ways forward in a majoritarian state. *The Caravan*. https://caravanmagazine.in/media/boiled-frog-indian-media-ways-forward-majoritarian-state

Kang, C., & Vogel, K. P. (2019). Tech giants amass a lobbying army for an epic Washington battle. *New York Times*. https://www.nytimes.com/2019/06/05/us/politics/amazon-apple-facebook-google-lobbying.html

Katta, S., Howson, K., Ustek-Spilda, F., & Graham, M. (2020). Uber and Deliveroo's 'charter of good work' is nothing but fairwashing. *Open Democracy*. https://www.opendemocracy.net/en/oureconomy/uber-and-deliveroos-charter-of-good-work-is-nothing-but-fairwashing/

Khabar Lahariya. (2020). Khabar Lahariya. http://khabarlahariya.org/

Khan, M., & Roy, P. (2019). Digital identities: A politcal settlements analysis of asymmetric power and information. *ACE SOAS Consortium Working Paper 015*. https://ace.soas.ac.uk/publication/digital-identities-a-political-settlements-analysis-of-asymmetric-power-and-information/

Khera, R. (2017). Impact of Aadhaar in welfare programmes. *Economic and Political Weekly*, *52*(5).

Khera, R. (2018). Aadhaar, welfare, and the media. *T.G. Narayanan Memorial Lecture*. https://www.youtube.com/watch?v=HiPSJio7jJY

Kidd, D. (2009). The Global Independent Media Center Network. In D. Mathison (Ed.), *Be the media*. Natural E Creative Group.

King, D. L., Case, C. J., & Premo, K. M. (2014). 2012 mission statements: A ten country global analysis. *Electronic Business Journal*, *13*(10).

King, G., Schneer, B., & White, A. (2017). How the news media activate public expression and influence national agendas. *Science*, *358*(6364), 776–780.

King, Jr, M.L. (1954). Rediscovering lost values. *Detroit, Michigan (USA)*. https://kinginstitute.stanford.edu/king-papers/documents/rediscovering-lost-values-0

Kinney, D. (2021). The mathematical case against blaming people for their misfortune. *Psyche*. https://psyche.co/ideas/the-mathematical-case-against-blaming-people-for-their-misfortune

Klapper, J. T. (1957). What we know about the effects of mass communication: The brink of hope. *The Public Opinion Quarterly*, *21*(4).

Kleine, D. (2010). ICT4WHAT? – Using the choice framework to operationalize the capability approach to development. *Journal of International Development*, *22*, 674–692.

Kocka, J. (1985). Marxist social analysis and the problem of white-collar employees. *State, Culture, and Society*, *1*(2).

Koradia, Z., Aggarwal, P., Seth, A., & Luthra, G. (2013). Gurgaon Idol: A singing competition over community radio and IVRS. *Proceedings of the 3rd ACM Symposium on Computing for Development, ACM DEV*.

Koradia, Z., Balachandran, C., Dadheech, K., Shivam, M., & Seth, A. (2012). Experiences of deploying and commercializing a community radio automation system in India. *Proceedings of the 2nd ACM Symposium on Computing for Development, ACM DEV*.

Korinek, A., Schindler, M., & Stiglitz, J. (2021). Technological progress, artificial intelligence, and inclusive growth. *IMF Working Paper*. https://www.imf.org/en/Publications/WP/Issues/2021/06/11/Technological-Progress-Artificial-Intelligence-and-Inclusive-Growth-460695

Kroes, P., Franssen, M., van de Poel, I., & Ottens, M. (2006). Treating socio-technical systems as engineering systems: Some conceptual problems. *Behavioral Science*, *23*(6).

Kulwin, N. (2018). The Internet apologizes. *The Intelligencer*. https://nymag.com/intelligencer/2018/04/an-apology-for-the-internet-from-the-people-who-built-it.html

Kumar, N. (2014). Facebook for self-empowerment? A study of Facebook adoption in urban India. *New Media Society*, *16*(7), 1122–1137.

Kumar, N. (2015). The gender-technology divide or perceptions of non-use? *First Monday*, *20*(11).

Kumar, S. (2021). The public opinion on farm laws has turned. *Mint*. https://www.livemint.com/news/india/why-the-bjp-may-choose-to-bury-the-controversial-farm-laws-now-11620626484918.html

Lam, M. (2014). Omlet: A revolution against big-brother social networks. *Proceedings of the 22nd ACM SIGSOFT International Symposium on Foundations of Software, FSE*.

Lampe, C., & Resnik, P. (2004). Slash(dot) and burn: Distributed moderation in a large online conversation space. *Proceedings of the 2004 ACM Conference on Human Factors in Computing Systems, ACM CHI*.

Larus, J., & Hankin, C. (2021). Regulating automated decision making. *Communications of the ACM*, *61*(8).

Latour, B. (1996). On actor–network theory: A few clarifications. *Soziale Welt*, *47*(4), 369–381

Ledger of Harms. (2020). Ledger of Harms. https://ledger.humanetech.com/

Lee, M. K., Kusbit, D., Kahng, A., Kim, J. T., Yuan, X., Chan, A., See, D., Noothigattu, R., Lee, S., Psomas, A., & Procaccia, A. D. (2018). WeBuildAI: Participatory frameworks for fair and efficient algorithmic governance. *Proceedings of the 2018 ACM Conference on Human Factors in Computing Systems, ACM CHI*.

Leeper, T. J., & Slothuus, R. (2018). Deliberation and framing. In A. Bächtiger, J. S. Dryzek, J. Mansbridge, & M. Warren (Eds.), *Oxford handbook of deliberative democracy*. Oxford University Press.

Lees, P. (2014). Facebook's gender identities are a good start – But why stop at 56? *The Guardian*. https://www.theguardian.com/commentisfree/2014/feb/14/facebook-gender-identity-56-transgender-cis

Ethical Source Licenses. (2020). The Anti-Capitalist Software License. https://anticapitalist.software/

Lim, M., Trere, E., Duran, O., Rodriguez, C., & Martinez, M. P. (2018). Roots, routes, routers: Communications and media of contemporary social movements. *Journalism and Communication Monographs*, *20*(2).

Little, D. (n.d.). A brief explanation of Marx's conception of false consciousness. *University of Michigan-Dearborn*. http://www-personal.umd.umich.edu/~delittle/iess%5C%20false%5C%20consciousness%5C%20V2.htm

Liu, W. (2018). Freedom isn't free: What would it take to set software free? *Logic*. https://logicmag.io/failure/freedom-isnt-free/

Lodato, T., & DiSalvo, C. (2018). Institutional constraints: The forms and limits of participatory design in the public realm. *Participatory Design Conference*.

Lupia, A., & Sen, G. (2003). Which public goods are endangered? How evolving communication technologies affect *The Logic of Collective Action*. *Public Choice*, *117*, 315–331.

Madaan, L., Sharma, A., Khandelwal, P., Goel, S., Singla, P., & Seth, A. (2019). Price forecasting and anomaly detection for agricultural commodities in India. *Proceedings of the 2nd ACM SIGCAS Conference on Computing and Sustainable Societies, ACM COMPASS*.

Mahajan, V., & Navin, T. (2013). Microfinance in India: Lessons from the Andhra crisis. In D. Köhn (Eds.), *Microfinance 3.0: Reconciling sustainability with social outreach and responsible delivery*. Springer Nature.

Mahfoud, T., Christine, A., Datta, S., & Rose, N. (2018). The limits of dual use. *Issues in Science and Technology*, *34*(4).

Manders-Huits, N. (2011). What values in design? The challenge of incorporating moral values into design. *Science and Engineering Ethics*, *17*(2), 271–287.

Mantelero, A. (2016). Personal data for decisional purposes in the age of analytics: From an individual to a collective dimension of data protection. *Computer, Law, and Security Review*, *32*(2).

Margalit, Y., & Shayo, M. (2020). How markets shape values and political preferences: A field experiment. *American Journal of Political Science*, *65*(2).

Marsden, G., Maunder, A., & Parker, M. (2008). People are people, but technology is not technology. *Philosophical Transactions of the Royal Society of London: Mathematical, Physical and Engineering Sciences*, *366*(1881), 3795–3804.

Stahl, B.C., & Goujon, P. (2010). Identifying the ethics of emerging information and communication technologies: An essay on issues, concepts and method. *International Journal of Technoethics*, *1*(4), 61–79.

Martin, C. J., Upham, P., & Klapper, R. (2017). Democratizing platform governance in the sharing economy: An analytical framework and initial empirical insights. *Journal of Cleaner Production*, *166*, 1395–1406.

Martinuzzi, A., Blok, V., Brem, A., & Stahl, B. C. (2018). Responsible research and innovation in industry – Challenges, insights and perspectives. *Sustainability*, *10*.

Karl Marx. (1880). A workers' inquiry. https://www.marxists.org/history/etol/newspape/ni/vol04/no12/marx.htm

Masiero, S. (2018). Subaltern studies: Advancing critical theory in ICT4D. *Research Papers*, *162*. https://aisel.aisnet.org/ecis2018_rp/162

Masiero, S., & Arvidsson, V. (2021). Degenerative outcomes of digital identity platforms for development. *Information Systems Journal*, *31*,

Masiero, S., & Bailur, S. (2021). Digital identity for development: The quest for justice and a research agenda. *Information Technology for Development*, *27*.

Matakos, A., Aslay, C., Galbrun, E., & Gionis, A. (2020). Maximizing the diversity of exposure in a social network. *IEEE Transactions on Knowledge and Data Engineering*.

Mathews, H. V. (2016). Flaws in the UIDAI process. *Economic and Political Weekly*, *51*(9).

Matias, J. N., & Mou, M. (2018). CivilServant: Community-led Experiments in platform governance. *Proceedings of the 2018 ACM Conference on Human Factors in Computing Systems, ACM CHI*.

Matson, E. W. (2020). A brief history of the editions of the theory of moral sentiments. *Liberty Fund, Adam Smith Works*. https://www.adamsmithworks. org/documents/a-brief-history-of-the-editions-of-tms-part-2

McCombs, M. E., & Shaw, D. L. (1972). The agenda-setting function of mass media. *The Public Opinion Quarterly*, *36*(2).

McGuigan, L. (2015). Proctor and Gamble, mass media, and the making of American life. *Media, Culture and Society*, *37*(6), 887–903.

McKie, R. (2021). Child labour, toxic leaks: The price we could pay for a greener future. *The Guardian*. https://www.theguardian.com/environment/2021/jan/03/child-labour-toxic-leaks-the-price-we-could-pay-for-a-greener-future

McPherson, M., Smith-Lovin, A., & Cook, J. (2001). Birds of a feather: Homophily in social networks. *Annual Review of Sociology*, *27*.

Mehra, A. (2021). Rote learning and the destruction of creativity. *The India Forum*. https://www.theindiaforum.in/article/rote-learning-school-snuffs-out-creativity

Menendez, A. J. (2000). Constituting deliberative democracy. *International Journal of Jurisprudence and Philosophy of Law*, *13*(4).

Merton, R. K. (1936). The unanticipated consequences of purposive social action. *American Sociological Review*, *1*(6).

Miller, K. (2013). A secret socio-technical system. *ITPro*, *15*(4).

Mills, R. (2011). Researching social news: Is Reddit.com a mouthpiece for the 'Hive Mind', or a collective intelligence approach to information overload? *ETHICOMP*.

Mocanu, D., Rossi, L., Zhang, Q., Karsai, M., & Quattrociocchi, W. (2015). Collective attention in the Age of (Mis) Information. *Computers in Human Behavior*.

Mohan, R. (2018). In Ranchi's Nagri Block, ration rice comes at a heavy price. *The Hindu*. https://www.thehindu.com/society/in-ranchis-nagri-block-ration-rice-comes-at-a-heavy-price/article23000824.ece

Moitra, A., Das, V., Gram Vaani, Kumar, A., & Seth, A. (2016). Design lessons from creating a mobile-based community media platform in rural India. *Proceedings of the International Conference of Information and Communication Technologies and Development (ICTD)*.

Moitra, A., Kumar, A., & Seth, A. (2018). An analysis of community mobilization strategies of a voice-based community media platform in rural India. *Information Technologies International Development*, 14.

Moitra, A., Kumar, A., & Seth, A. (2019). An analysis of impact pathways arising from a mobile-based community media platform in rural India. *Working Paper*. https://arxiv.org/abs/2104.07901

Mondal, M., Silva, L. A., & Benevenuto, F. (2017). A measurement study of hate speech in social media. *Proceedings of ACM Hypertext and Social Media Conference*.

Moor, J. H. (1985). What is computer ethics? *Metaphilosophy*, *16*(4).

Moor, J. H. (2006). The nature, importance, and difficulty of machine ethics. *IEEE Intelligent Systems*.

Moore, A., & Himma, K. (2018). Intellectual property. *Stanford Encyclopedia of Philosophy*. https://plato.stanford.edu/entries/intellectual-property/

Moore, J. (2020). Towards a more representative politics in the ethics of computer science. *Proceedings of the 2020 ACM Conference on Fairness, Accountability, and Transparency*.

Morduch, J., & Haley, B. (2002). Analysis of the effects of microfinance on poverty reduction. *NYU Wagner Working Paper No. 1014*. https://pdf.wri.org/ref/morduch_02_analysis_effects.pdf

Mosse, D. (2018). Caste and development: Contemporary perspectives on a structure of discrimination and advantage. *World Development*, *110*, 422–436.

Mudliar, P., Donner, J., & Thies, W. (2013). Emergent practices around CGNet Swara, a voice forum for citizen journalism in rural India. *Information Technologies and International Development*, *9*(2).

Mullainathan, S., & Shleifer, A. (2005). The market for news. *American Economic Review, 95*(1).

Mulvenna, M., Boger, J., & Bond, R. (2017). Ethical by design – A manifesto. *Proceedings of the European Conference on Cognitive Ergonomics, ACM ECCE.*

Munson, S. A., Lee, S. Y., & Resnick, P. (2013). Encouraging reading of diverse political viewpoints with a browser widget. *Proceedings of the International AAAI Conference on Web and Social Media, ICWSM.*

Muskaan, Dhaliwal, M. P., & Seth, A. (2019). Fairness and diversity in the recommendation and ranking of participatory media content. *ACM SIGKDD International Conference on Knowledge Discovery and Data Mining, Workshop on Intelligent Information Feeds*. https://arxiv.org/abs/1907.07253

Nadler, A., & McGuigan, L. (2018). An impulse to exploit: The behaviorial turn in data-driven marketing. *Critical Studies in Media Communication, 35*(2).

Narayanan, A. (2018). 21 Fairness definitions and their politics. *Tutorials at FAT**. https://www.youtube.com/watch?v=jIXIuYdnyyk

Narayanan, S. (2020). The three farm bills: Is this the market reform Indian agriculture needs? *The India Forum*. https://www.theindiaforum.in/article/three-farm-bills

Navarro, P. (2001). The limits of social conversation: A sociological approach to Gordon Pask's Conversation Theory. *Kybernetes, 30*(5).

Neelakantan, M., Barooah, S., Dasgupta, S., & Sarkar, T. (2018). National Health Stack: Data for data's sake, a manmade health hazard. *Bloomberg Quint*. https://www.bloombergquint.com/opinion/data-for-datas-sake-a-manmade-health-hazard

Neelima, M. S. (2020). Is India privatizing governance through partnerships in public digital infrastructure? *The Caravan*. https://caravanmagazine.in/policy/is-india-privatising-governance-through-partnerships-public-digital-infrastructure

Neyland, D. (2016). Bearing accountable witness to the ethical algorithmic system. *Science, Technology and Human Values, 4*(1).

Nikolov, D., Flammini, A., & Menczer, F. (2021). Right and left, partisanship predicts (asymmetric) vulnerability to misinformation. *Harvard Kennedy School Misinformation Review*.

Nilekani, N. (2017). Societal platforms. *Center for Global Development*. https://www.youtube.com/watch?v=oiGmBnb94uM

Nissenbaum, H. (1996). Accountability in a computerized society. *Science and Engineering Ethics, 2*, 25–42.

Nissenbaum, H. (2011). From preemption to circumvention: If technology regulates, why do we need regulation (and vice versa). *Berkeley Technology Law Journal, 26*(3).

Nohria, N., & Taneja, H. (2021). Managing the unintended consequences of your innovations. *Harvard Business Review*. https://hbr.org/2021/01/managing-the-unintended-consequences-of-your-innovations

NYT Editorial. (2021). What happens when you click 'agree'? *New York Times*. https://www.nytimes.com/2021/01/23/opinion/sunday/online-terms-of-service.html

Oquendo, A. R. (2002). Deliberative democracy in Habermas and Nino. *Oxford Journal of Legal Studies, 22*(2), 189–226.

Owen, C., Todor, W. D., & Dalton, D. R. (1989). A comparison of blue-versus white-collar grievance behavior: A field assessment. *Employee Responsibilities and Rights Journal, 2*(4).

P2P Foundation. (2019). 78 Questions to ask about any technology. *P2P Foundation*. https://blog.p2pfoundation.net/78-questions-to-ask-about-any-technology/2019/02/26

Pal, J. (2008). Computers and the promise of development: Aspiration, neoliberalism and 'technolity' in India's ICTD Enterprise. *Confronting the challenge of technology for development: Experiences from the BRICS.*

Pal, R., Nag, B., Crowcroft, J., Liu, M., Ghosh, P., & De, S. (2021). Fixing the data economy, and economic inequality. *Financial express*. https://www.financialexpress.com/opinion/fixing-the-data-economy-and-economic-inequality/2356761/

Papadimitropoulos, V. (2017). The politics of the commons: Reform or revolt? *tripleC: Communication, Capitalism, and Critique*, *15*(2).

Papadogiannakis, E., Papadopoulos, P., Kourtellis, N., & Markatos, E. P. (2021). User tracking in the post-cookie Era: How websites bypass GDPR consent to track users. *Proceedings of The Web Conference.*

Park, S., Kang, S., Chung, S., & Song, J. (2009). NewsCube: Deliversing multiple aspects of news to mitigate media bias. *Proceedings of the 2009 ACM Conference on Human Factors in Computing Systems, ACM CHI.*

Patel, N., Chittamuru, D., Jain, A., Dave, P., & Parikh, T. S. (2010). Avaaj Otalo – A field study of an interactive voice forum for small farmers in rural India. *Proceedings of the 2010 ACM Conference on Human Factors in Computing Systems, ACM CHI.*

Pedersen, J. (2015). War and peace in codesign. *International Journal of CoCreation in Design and the Arts*, *12*(3), 171–184.

Pettit, J. (2013). Power analysis: A practical guide. *SIDA*. https://usaidlearninglab.org/library/power-analysis-practical-guide

Pfeffer, J. (1992). Understanding power in organizations. *California Management Review*, *34*(2), 29–51.

Phansalkar, S. (2020). Of cockroaches and digital health records. *VillageSquare*. https://www.villagesquare.in/2020/09/02/of-cockroaches-and-digital-health-records/

Planning Commission. (2019). Frequently asked questions about Aadhaar. What is the utility of the Aadhaar number? *Planning Commission, Government of India*. http://planningcommission.nic.in/sectors/dbt/faq_dbt220313.doc

Poltrock, S., & Grudin, J. (1994). Computer supported cooperative work and groupware. *Proceedings of the 1994 ACM Conference on Human Factors in Computing Systems, ACM CHI.*

Prado, J. (2018). Prospects for organizing the tech industry. *Technology and the Worker*. https://notesfrombelow.org/article/prospects-for-organizing-the-tech-industry

Prat, A., & Stromberg, D. (2013). The political economy of mass media. In D. Acemoglu, M. Arellano, & E. Dekel (Eds.), *Advances in economics and econometrics* (pp. 135–187). Cambridge University Press.

Prior, M. (2001). Efficient choice, inefficient democracy? The implications of cable and internet access for political knowledge. *TPRC conference*.

Radhakrishnan, S. (2015). "Low Profile" or entrepreneurial? Gender, class, and cultural adaptation in the global microfinance industry. *World Development*, *74*, 264–274.

Ramanathan, U. (2017). Without Supreme Court interference, the Aadhaar Project is a ticking time bomb. *The Wire*. https://thewire.in/government/aadhaar-supreme-court-uid

Russell Einstein Manifesto. (1955). Russell–Einstein Manifesto. https://www.atomicheritage.org/key-documents/russell-einstein-manifesto

Ribeiro, M. T., Singh, S., & Guestrin, C. (2016). "Why should I trust you?": Explaining the predictions of any classifier. *Proceedings of the 2016 ACM SIGKDD International Conference on Knowledge Discovery and Data Mining*.

Rittel, H. W. J., & Webber, M. M. (1973). Dilemmas in a general theory of planning. *Policy Sciences*, *4*.

de Rivera, J. (2020). A guide to understanding and combatting digital capitalism. *tripleC: Communication, Capitalism, and Critique*, *18*(2).

Rodriguez, C., Ferron, B., & Shamas, K. (2014). Four challenges in the field of alternative, radical and citizens' media research. *Media, Culture, and Society*, *17*, 150–166.

Rodriguez-Labajos, B., Yanez, I., Bond, P., Greyl, L., Munguti, S., Ojo, G. U., & Overbeek, W. (2019). Not so natural an alliance? Degrowth and environmental justice movements in the Global South. *Ecological Economics*, *157*.

Rogaway, P. (2015). The moral character of cryptographic work. *Cryptology ePrint Archive, Report 2015/1162*. https://ia.cr/2015/1162.

Rogerson, S. (2010). A review of information ethics. *Japan Society for Information and Management*, *30*(3).

Rogerson, S. (2015). Future vision. *Journal of Information Communication and Ethics in Society*, *13*(3).

Rogerson, S. (2017). Coding ethics into technology. *Hack Craft News*, July 3, Additional reporting by Chris Middleton. http://hncnews.com/coding-ethics-technology

Rojas, H., & Puig-i-Abril, E. (2009). Mobilizers mobilized: Information, expression, mobilization and participation in the Digital Age. *Journal of Computer-Mediated Communication*, *14*, 902–927.

Roose, K. (2019). The making of a Youtube radical. *New York Times*. https://www.nytimes.com/interactive/2019/06/08/technology/youtube-radical.html

Roose, K. (2021). On Election Day, Facebook and Twitter did better by making their products worse. *New York Times*. https://www.nytimes.com/2020/11/05/technology/facebook-twitter-election.html

Rose, S., & Rose, H. (1973). Can science by neutral? *Perspectives in Biology and Medicine*, *16*(4).

Rosenblat, A., & Stark, L. (2016). Algorithmic labour and information asymmetries: A case-study of Uber drivers. *International Journal of Communication*, *10*, 3758–3784.

Roy, I., Ajmal, Z., Anand, A., Jaiswal, A., Raman, R., Gupta, O., Sawant, S., Pandey, J., Gupta, R., Shetty, T., Sharma, G., & Prajapati, R. (2021). Precarious transitions: Mobility and citizenship in a rising power. *Economic and Political Weekly*, *56*(7). https://www.epw.in/engage/article/precarious-transitions-mobility-and-citizenship

Ruthven, O. (2020). Lockdown Chronicle: The story of a migrant workers' platform across India's lockdown. *The Journal of Agrarian Change*. https://www.aqs.org.uk/lockdown-chronicle-the-story-of-a-migrant-workers-platform-across-indias-lockdown/

Sachitanand, R. (2018). Voice, video and vernacular: India's Internet Landscape is changing to tap new users. *The Economic Times*. https://bit.ly/35iYyyJ

Sanders, E. B., & Stappers, P. J. (2008). Co-creation and the New Landscapes of design. *CoDesign*, *4*(1), 5–18.

Sanders, L. (2008). An evolving map of design practice and design research. *ACM Interactions*, *15*(6). https://doi.org/10.1145/1409040.1409043

Sandoval, M. (2019). Entrepreneurial activism? Platform cooperativism between subversion and co-optation. *Critical Sociology*, *46*(6), 801–817.

Saurwein, F., Just, N., & Latzer, M. (2015). Governance of algorithms: Options and limitations. *info*, *17*(6), 35–49.

Saveski, M., Gillani, N., Yuan, A., Vijayaraghavan, P., & Roy, D. (2021). Perspective-taking to reduce affective polarization on social media. *Proceedings of AAAI*.

Sawhney, S. (2021). Res-Publica: The ground we share. *The Wire*. https://thewire.in/rights/republic-day-india-protest-farmers

Scannell, J. (2015). What can an algorithm do? *DIS Magazine*. http://dismagazine.com/discussion/72975/josh-scannell-what-can-an-algorithm-do/

Scheufele, D. A., & Tewksbury, D. (2007). Framing, agenda setting, and priming: The evolution of three media effects models. *Journal of Communication*, *57*.

Schiffer, E. (2006). The power mapping tool: A method for the empirical research of power relations. *IFPRI Discussion Paper 00703. International Food Policy Research Institute*. https://ideas.repec.org/p/fpr/ifprid/703.html

Schmalz, S., Ludwig, C., & Webster, E. (2018). The power resources approach: Developments and challenges. *Global Labour Journal*, *9*(2).

Schneider, C., Weinmann, M., & Brocke, J. (2018). Digital nudging – Influencing choices by user interface design. *Communications of the ACM*, *61*(7), 67–73.

Schneider, H., Eibandm M., Ullrich, D., & Butz, A. (2018). Empowerment in HCI: A survey and framework. *Proceedings of the 2018 ACM Conference on Human Factors in Computing Systems, ACM CHI*.

Scholz, T. (2016). *Platform cooperativism: Challenging the corporate sharing economy*. Rosa Luxemburg Stiftung. https://rosalux.nyc/wp-content/uploads/2020/11/RLS-NYC_platformcoop.pdf

SDGs. (2015). Social development goals. *United Nations*. https://www.undp.org/content/undp/en/home/sustainable-development-goals.html

Sen, A. (2021). A computer-based approach to analyze some aspects of the political economy of policy making in India. *Ph.D. thesis, Indian Institute of Technology Delhi.*

Sen, A., Agarwal, A., Guru, A., Choudhuri, A., Singh, G., Mohammed, I., Goyal, J., Mittal, K., Singh, M., Goel, M., & Seth, A. (2018). Leveraging web data to monitor changes in corporate–government interlocks in India. *Proceedings of the ACM SIGCAS conference on computing and sustainable societies*, ACM COMPASS.

Sen, A., Ghatak, D., Kumar, K., Khanuja, G., Bansal, D., Gupta, M., Rekha, K., Bhogale, S., Trivedi, P., & Seth, A. (2019). Studying the discourse on economic policies in India using mass media, social media, and the parliamentary question hour data. *Proceedings of the ACM SIGCAS conference on computing and sustainable societies*. ACM COMPASS

Sen, D., Priya, Aggarwal, P., Verma, S., Ghatak, D., Kumari, P., Singh, M., Guru, A., Seth, A. (2019). An attempt at using mass media data to analyze the political economy around some key ICTD policies in India. *Proceedings of the 2019 Tenth International Conference on Information and Communication Technologies and Development (ICTD)*.

New School Series. (2020). Can companies deliver equity? *Management and Social Justice Conversation Series*, The New School. https://www.youtube.com/watch?v=5l4svQuYZWQ&feature=youtu.be

Seth, A. (2008). Design of a recommender system for participatory media built on a tetherless communication infrastructure. *Ph.D. thesis, University of Waterloo.*

Seth, A. (2019a). A new paradigm to accommodate ethical foundations in the design and management of digital platforms. *Manuscript, Indian Institute of Technology Delhi.* http://www.cse.iitd.ernet.in/~aseth/systems-thinking-2019.pdf

Seth, A. (2019b). Ensuring responsible outcomes from technology. *Proceedings of IEEE International Conference on Communication Systems and Networks and Workshops, COMSNETS.*

Seth, A. (2020a). Ethics in applied computer science. *Course contents, Indian Institute of Technology Delhi.* http://bit.ly/2FARXrs

Seth, A. (2020b). Learning to listen: Building an empathetic state. *The India Forum*. https://www.theindiaforum.in/article/learning-listen

Seth, A. (2020c). Technologies that disempower: Design flaws in technologies used for the delivery of social protection measures in India. *ACM Interactions*, *27*(6).

Seth, A. (2020d). The elusive model of technology, media, social development, and financial sustainability. In L. Poonamallee, J. Scillitoe, & S. Joy (Eds.), *Harnessing technology development for social impact*. Palgrave MacMillan.

Seth, A. (2020e). The limits of design in ensuring responsible outcomes from technology. *Proceedings of the 2020 International Conference on Information and Communication Technologies and Development, ICTD.*

Seth, A. (2021a). A call to technologists to ensure that responsible outcomes arise from their innovations. *Journal of Information, Communication and Ethics in Society*, *19*(2) 268–279.

Seth, A. (2021b). The Post-lockdown state of labour. *India Development Review*. https://idronline.org/labour-rights-in-india-have-worsened-post-lockdown-covid-19/

Seth, A., Ahmad, S., & Ruthven, O. (2020). NotStatusQuo: A campaign to fix the broken social protection system in India. *Gram Vaani*. https://www.cse.iitd.ac.in/~aseth/notstatusquo-final-report.pdf

Seth, A., Gupta, A., Moitra, A., Kumar, D., Chakraborty, D., Enoch, L., Ruthven, O., Panjal, P., Siddiqi, R. A., Singh, R., Chatterjee, S., Saini, S., & Ahmad, S. Pratap, S. V. (2020). Reflections from practical experiences of managing participatory media platforms for development. *Proceedings of the 2020 International Conference on Information and Communication Technologies and Development, ICTD*.

Seth, A., & Zhang, J. (2008). A social network based approach to personalized recommendation of participatory media content. *Proceedings of the International AAAI Conference on Web and Social Media, ICWSM*.

Seth, A., Zhang, J., & Cohen, R. (2015). A personalized credibility model for recommending messages in social participatory media environments. *World Wide Web, 18*(1), 111–137.

Shah, D. V., Cho, J., Nah, S., Gotlieb, M. R., Hwang, H., Lee, N., Scholl, R. M., McLeod, D. M. (2007). Campaign Ads, online messaging, and participation: Extending the communication mediation model. *Journal of Communication, 57*.

Shah, D. V., McLeod, D. M., Rojas, H., Cho, J., Wagner, M. W., & Friedland, L. A. (2017). Revising the communication mediation model for a new political communication ecology. *Human Communication Research, 43*.

Shane, S., & Wakabayashi, D. (2018). 'The business of war': Google employees protest work for the Pentagon. *New York Times*. https://www.nytimes.com/2018/04/04/technology/google-letter-ceo-pentagon-project.html

Sharma, A., Kaur, N., Sen, A., & Seth, A. (2020). Ideology detection in the Indian mass media. *IEEE/ACM International Conference on Advances in Social Networks Analysis and Mining, ASONAM*.

Shashidhar, K. J. (2018). Know your creator: The evolution of KYC from checkbox to disruptor. *The Ken*. https://the-ken.com/story/kyc-business-disruption/

Shilton, K., Koepfler, J. A., & Fleischmann, K. R. (2013). Charting sociotechnical dimensions of values for design research. *The Information Society, 29*(5).

Shneiderman, B. (1990). Human values and the future of technology: A declaration of empowerment. *Keynote address, ACM SIGCAS conference: Computers and the Quality of Life*.

Shneiderman, B., & Rose, A. (1996). Social impact statements: Engaging public participation in information technology design. *ACM Conference on Computers and the Quality of Life, CQL*.

Simons, J. (2019). The politics of machine learning: Discrimination, fairness, and equality. *Personal communication*.

Singh, S. (2021). Inside BCG's Long Game to defeat McKinsey in India. *The Ken*, https://the-ken.com/story/inside-bcg-long-game-to-defeat-mckinsey-in-india/

Singh, S., & Ramanathan, A. (2020). The Elite VC-founder Club Riding Aarogya Setu to Telemed Domination. *The Ken*. https://the-ken.com/story/the-elite-vc-founder-club-riding-aarogya-setu-to-telemed-domination/

Smith, A. (2014). The Lucas Plan: What can it tell us about democratising technology today? *The Guardian*. https://www.theguardian.com/science/political-science/2014/jan/22/remembering-the-lucas-plan-what-can-it-tell-us-about-democratising-technology-today

Smith, K. T., & Alexander, J. J. (2013). Which CSR-related headings do Fortune 500 companies use on their websites? *Business Communication Quarterly, 76*(2).

Smyth, T. N., Kumar, S., Medhi, I., & Toyama, K. (2010). Where there's a will there's a way: Mobile media sharing in Urban India. *Proceedings of the 2010 ACM Conference on Human Factors in Computing Systems, ACM CHI*.

Sood, A. (2020). The silent takeover of labour rights. *The India Forum*. https://www.theindiaforum.in/article/silent-takeover-labour-rights

Sovacool, B. K., Turnheim, B., Hook, A., Brock, A., & Martiskainen, M. (2021). Dispossessed by decarbonization: Reducing vulnerability, injustice, and inequality in the lived experience of low-carbon pathways. *World Development, 137*.

Spinuzzi, C. (2005). The methodology of participatory design. *Technical Communication, 52*(2), 163–174.

Spivak, G. C. (1988). Can the Subaltern speak? In C. Nelson & L. Grossberg (Eds.), *Marxism and the interpretation of culture* (pp. 271–313). Macmillan Education.

Srivas, A. (2021). Understanding the nuances to Twitter's Standoff with the Modi Government. *The Wire*. https://thewire.in/tech/twitter-modi-government-block-section-69-a

Stallman, R. (2007). Why open source misses the point of free software? *Free Software Foundation*. https://www.gnu.org/philosophy/open-source-misses-the-point.en.html

Stallman, R. (2013). How much surveillance can democracy withstand? *Wired*. https://www.wired.com/2013/10/a-necessary-evil-what-it-takes-for-democracy-to-survive-surveillance/

Stark, L., & Hoffmann, A. L. (2019). Data is the new what? Popular metaphors and professional ethics in emerging data culture. *Journal of Cultural Analytics, 4*(1).

Stoddard, G. (2017). Popularity dynamics and intrinsic quality in Reddit and Hacker News. *Proceedings of the International AAAI Conference on Web and Social Media, ICWSM*.

Suchman, L. (2002). Located accountabilities in technology production. *Scandinavian Journal of Information Systems, 14*(2), 91–105.

Surana, S., Patra, R., Nedevschi, S., Ramos, M., Subramanian, L., Ben-david, Y., & Brewer, E. (2008). Beyond pilots: Keeping rural wireless networks alive. *USENIX Symposium on Networked Systems Design and Implementation, NSDI*.

Surowiecki, J. (2012). Unequal shares. *New Yorker*. https://www.newyorker.com/magazine/2012/05/28/unequal-shares

Syreng, T. A., & Boyd, F. J. (2020). Employer and employee ownership of intellectual property: Not as easy as you think. *Thomson Reuters*. https://store.legal.thomsonreuters.com/law-products/news-views/corporate-counsel/employer-and-employee-ownership-of-intellectual-property-not-as-easy-as-you-think

Tacchi, J. (2012). Digital engagement: Voice and participation in development. In H. A. Horst & D. Miller (Eds.), *Digital anthropology*. Routledge.

Tavani, H. (2007). Philosophical theories of privacy: Implications for an adequate online privacy policy. *Metaphilosophy, 38*(1).

Theuws, M., & Overeem, P. (2014). Flawed fabrics: The abuse of girls and women workers in the South Indian textile industry. *Center for Research on Multinational Corporations, India Committee of the Netherlands*. https://www.somo.nl/flawed-fabric-the-abuse-of-girls-and-women-workers-in-the-south-indian-textile-industry/

Thompson, N., & Vogelstein, F. (2018). Inside the two years that shook Facebook – And the world. *Wired Magazine*. https://www.wired.com/story/inside-facebook-mark-zuckerberg-2-years-of-hell/

Tiwari, P. (2021). The high cost of India's cheap garment exports. *Al Jazeera*. https://www.aljazeera.com/opinions/2021/4/19/the-high-cost-of-indias-cheap-garment-exports

Torlak, N. G., & Muceldili, B. (2014). Soft systems methodology in action: The example of a private hospital. *Systemic Practice and Action Research, 27*, 325–361.

Toyama, K. (2017). From needs to aspirations in information technology for development. *Information Technology for Development, 24*(1), 15–36.

Treem, J. W., Dailey, S. L., Pierce, C. S., & Biffl, D. (2016). What are we talking about when we talk about social media: A framework for study. *Sociology Compass, 10*(9), 768–784.

Tufekci, Z. (2016). The real bias built in at Facebook. *New York Times*. https://www.nytimes.com/2016/05/19/opinion/the-real-bias-built-in-at-facebook.html

Turchin, P. (2018). Does capitalism destroy cooperation. http://peterturchin.com/cliodynamica/does-capitalism-destroy-cooperation/

Turpin, M., & Alexander, P. M. T. (2014). Desperately seeking systems thinking in ICT4D. *Electronic Journal of Information Systems in Developing Countries, 61*(1).

UDHR. (1948). Universal declaration of human rights. *United Nations*. https://www.un.org/en/universal-declaration-human-rights/

Unwin, T. (2018a). Contributions to UNESCO's first Partners' Forum: Notes from the under-ground. https://unwin.wordpress.com/

Unwin, T. (2018b). Why we shouldn't use terms such as "bridging the digital divide" or "digital leapfrogging". https://unwin.wordpress.com/

Urquhart, C., & Andrade, A. D. (2012). Unveiling the modernity bias: A critical examination of the politics of ICT4D. *Information Technology for Development, 18*(4), 281–292.

Vaccaro, K., & Karahalios, K. (2018). Algorithmic appeals. *NSF Trustworthy Algorithmic Decision-Making Workshop*.

Vaidhyanathan, S. (2019). Dear Mr Zuckerberg: The problem isn't the internet, it's Facebook. *The Guardian*. https://www.theguardian.com/technology/2019/feb/04/facebook-15-anniversary-mark-zuckerberg

Valkenburg, P. M. (2017). Understanding self-effects in social media. *Human Communication Research, 43*.

Vallor, S., Green, B., & Raicu, I. (2018). Conceptual frameworks in technology and engineering practice: Ethical lenses to look through. *Ethics in Technology Practice, Markkula Center of Applied Ethics*.

Vandenberghe, B., & Slegers, K. (2016). Designing for others, and the trap of HCI methods and practices. *Proceedings of the 2016 ACM Conference on Human Factors in Computing Systems, ACM CHI*.

Vashistha, A., Cutrell, E., Borriello, G., & Thies, B. (2015). Sangeet Swara: A community-moderated voice forum in rural India. *Proceedings of the 2015 ACM Conference on Human Factors in Computing Systems, ACM CHI*.

Veeraraghavan, R. (2013). Dealing with the digital panopticon: The use and subversion of ICT in an Indian bureaucracy. *Proceedings of the 2013 International Conference on Information and Communication Technologies and Development, ICTD*.

Venkatanarayanan, A. (2017). The curious case of the World Bank and Aadhaar savings. *The Wire*. https://thewire.in/economy/the-curious-case-of-the-world-bank-and-aadhaar-savings

Verma, S., & Rubin, J. (2018). Fairness definitions explained. *ACM FairWare: Proceedings of the International Workshop on Software Fairness*.

Video Volunteers. (2020). Video volunteers. https://www.videovolunteers.org/

Vines, J., Clarke, R., Wright, P., McCarthy, J., & Olivier, P. (2013). Configuring participation: On how we involve people in design. *Proceedings of the 2013 ACM Conference on Human Factors in Computing Systems, ACM CHI*.

Vistisen, P., & Jensen, T. (2013). The ethics of user experience design. *ETHICOMP*.

Volkoss, O., & Strong, D. (2017). Affordance theory and how to use it in IS research. In R.D. Galliers & M. Stein (Eds.), *The Routledge companion to management information systems*. Routledge.

Vosoughi, S., Roy, D., & Aral, S. (2018). The spread of true and false news online. *Science, 359*(6380), 1146–1151.

Vredenburg, K., Mao, J., Smith, P. W., & Carey, T. (2002). A survey of user-centered design practice. *Proceedings of the 2002 ACM Conference on Human Factors in Computing Systems, ACM CHI*.

de Vries, K. (2010). Identity, profiling algorithms and a world of ambient intelligence. *Ethics and Information Technology, 12*.

Wade, R. H. (2002). Bridging the digital divide: New route to development or new form of dependency? *Global Governance, 8*(4).

Wade, R. H. (2004). ICT, power, and developmental discourse: A critical analysis. *Electronic Journal on Information Systems in Developing Countries, 20*(4).

Wagner, B. (2018). Ethics as an escape from regulation: From ethics-washing to ethicsshopping. In M. Hildebrandt (Ed.), *Being profiling*. Amsterdam University Press.

Wagner, B. (2019). Liable, but not in control? Ensuring meaningful human agency in automated decision-making systems. *Policy and Internet, 11*(1).

Waldman, A. E. (2018). Designing without privacy. *Houston Law Review, 55*(659).

Wang, H. H., Seth, A., Johri, M., Kalra, E., & Singhal, A. (2021). Communication infrastructure and community mobilization: The case of Gram Vaani's COVID-19 response network for the marginalized in India. *Journal of Development Communication, 32*(2).

Washington, A. L., & Kuo, R. (2020). Whose side are ethics codes on? Power, responsibility and the social good. *Proceedings of the 2020 ACM Conference on Fairness, Accountability, and Transparency*.

White, K., Habib, R., & Hardisty, D. J. (2019). How to SHIFT consumer behaviors to be more sustainable: A literature review and guiding framework. *Journal of Marketing, 83*(3).

Wihbey, J. (2014). The challenges of democratizing news and information: Examining data on social media, viral patterns and digital influence. *Shorenstein Center of Media, Politics and Public Policy. Discussion Paper Series*. https://ssrn.com/abstract=2466058

Wikipedia. (2022). Whole earth catalog. *English Wikipedia*. https://en.wikipedia.org/wiki/Whole_Earth_Catalog

Wilder, C. (1997). Being analog. *International Communication Association*. http://cat4chat.narod.ru/wilder_analog.htm

Wilemme, G. (2018). Regulating Uber. *Paris Innovation Review*. http://parisinnovationreview.com/articles-en/regulating-uber

Williams, C., & Blaiklock, A. (2015). With SDGs now adopted, human rights must inform implementation and accountability. *Health and Human Rights Journal*. https://www.hhrjournal.org/2015/09/sdg-series-with-sdgs-now-adopted-human-rights-must-inform-implementation-and-accountability/

Winkler, T., & Spiekermann, S. (2018). Twenty years of value sensitive design: A review of methodological practices in VSD projects. *Ethics and Information Technology, 23*(1)

Winner, L. (1980). Do artefacts have politics. *Modern Technology: Problem or Opportunity? Daedalus, 109*(1), 121–136.

Wire. (2016). Right to privacy a fundamental right, says Supreme Court in unanimous verdict. *The Wire*. https://thewire.in/law/supreme-court-aadhaar-right-to-privacy.

Wong, J. C. (2017). Former Facebook executive: Social media is ripping society apart. *The Guardian*. https://www.theguardian.com/technology/2017/dec/11/facebook-former-executive-ripping-society-apart

Woodard, J., Cohen, C., Cox, C., Fritz, S., Johnson, D., Koo, J., Mclean, M., See, L., Speck, T., & Sturn, T. (2016). Using ICT for remote sensing, crowdsourcing, and big data to unlock the potential of agricultural data. *World Bank Group*. https://elibrary.worldbank.org/doi/10.1596/978-1-4648-1002-2_Module15

Woodcock, J. (2021). Towards a digital workerism: Workers' inquiry, methods, and technologies. *Nanoethics, 15*.

Wrobel, B., Basoya, S., & Menon, D. (2019). Catalyzing Civic Tech in India. *Omidyar Network*. https://www.omidyarnetwork.in/wp-content/uploads/2019/10/Catalyzing-Civic-Tech-10.15_compressed-1.pdf

Wu, T. (2018). The tyranny of convenience. *New York Times*. https://www.nytimes.com/2018/02/16/opinion/sunday/tyranny-convenience. html

Xiao, L., Farooq, U., Lee, R. L., & Carroll, J. M. (2004). Participatory design in community computing contexts. *Proceedings of the 2004 Conference on Participatory Design*.

Yadav, D., Singh, P., Montague, K., Kumar, V., Sood, D., Balaam, M., Sharma, D., Duggal, M., Bartindale, T., Varghese, D., & Olivier, P. (2017). Sangoshthi: Empowering community health workers through peer learning in rural India. *Proceedings of the 26th International Conference on World Wide Web, WWW*.

Young, T. R. (1985). The structure of democratic communications. *Mid-American Review of Sociology*, *10*(2), 55–76.

Zahra, S. A., Gedajlovic, E., Neubaum, D. O., & Shulman, J. M. (2009). A typology of social entrepreneurs: Motives, search processes and ethical challenges. *Journal of Business Venturing*, *24*, 519–532.

de Zeeuw, G. (2001). Interaction of actors theory. *Kybernetes*, *30*(7), 971–983.

Zuckerberg, M. (2019). 15 Years of Facebook. *Facebook post*. https://www.facebook.com/zuck/posts/10106411140260321

Index

Note: Page numbers followed by "*n*" indicate notes.